MARINES AT WAR

STORIES FROM AFGHANISTAN AND IRAQ

Edited by
Paolo G. Tripodi and Kelly Frushour

Foreword by
Carroll J. Connelley

Quantico, Virginia

Library of Congress Cataloging-in-Publication Data

Names: Tripodi, Paolo G., editor of compilation. | Frushour, Kelly, editor of
  compilation.
Title: Marines at war : stories from Afghanistan and Iraq / edited by Paolo
  G. Tripodi and Kelly Frushour.
Other titles: Stories from Afghanistan and Iraq
Description: Quantico, VA : Marine Corps University Press, [2016] | Includes
  bibliographical references.
Identifiers: LCCN 2016019036
Subjects: LCSH: Iraq War, 2003-2011--Personal narratives, American. | Afghan
  War, 2001---Personal narratives, American. | United States. Marine
  Corps--Biography. | Marines--United States--Biography.
Classification: LCC DS79.766.A1 M37 2016 | DDC 956.7044/345092273--dc23
LC record available at https://lccn.loc.gov/2016019036

Disclaimer

The views expressed in this publication are solely those of the authors. They do not necessarily reflect the opinions of the organizations for which they work, Marine Corps University, the U.S. Marine Corps, the Department of the Navy, or the U.S. government. The information contained in this book was accurate at the time of printing.

Published by
Marine Corps University Press
3078 Upshur Avenue
Quantico, VA
22134
www.mcu.usmc.mil/mcu_press

1st Printing 2016
ISBN 978-0-9973174-1-1

DEDICATION

In memory of Major Douglas A. Zembiec and all of the fallen and wounded U.S. Marines of the wars in Afghanistan and Iraq.

CONTENTS

FOREWORD

The decade plus of U.S. military combat in Afghanistan and Iraq has resulted in an extraordinary amount of articles, books, documentaries, and films. As with many subjects in U.S. culture today, the resulting productions are often glamorized, politicized, or simply focused on only the most astonishing events of these two conflicts. In contrast, this collection of narratives focuses on the more ordinary stories of war.

This book brings together the short stories of 10 Marines and a sailor who served in either Afghanistan or Iraq, some in both. The authors reflect on their time in combat, focusing on preparation, particular moments, and the lessons or conclusions drawn from their experiences. While every servicemember has a different story, and despite these remembrances being very personal, this collection represents and provides insight into the ordinary experiences of fellow Americans serving during an extraordinary set of circumstances. That set of circumstances is obviously war. War, for the modern American, is the monotony of everyday life lived far away from home, broken up by intense moments of fear, conflict, and violence. This collection examines both the monotony and the intense moments that these men and women experienced from a diverse set of perspectives. From the Bell AH-1 Cobra helicopter pilot to the logistician to the chaplain and beyond, each experience is different and informative in its own particular sphere.

I personally felt these stories were important to share because of my own experience in Iraq. As a former infantry officer turned judge advocate, I went to Iraq with the expectation of learning more of my craft as a relatively new lawyer and finally getting a chance to experience war. After a six-month deployment and upon returning home in 2006, I felt pride at having been part of the unit that made the al-Anbar Province more amenable to self-government and less hospitable to terrorists and insurgents. The regimental combat team I had been a member of for six months in fact had established conditions for what would later become known as the "al-Anbar Awakening," one of the shining moments of the Iraq War. However, that success was attributed to follow-on forces, and my unit's time in Iraq became inextricably linked to the incident at Haditha in which multiple civilians were killed, resulting in a political and judicial firestorm that continued for years. Although I was not directly involved in the incident, I was a part of the chain of command responsible for responding and reporting the incident. I have spent years considering my part in that response and what I might have done differently. The incident affected my career, but in a very minimal way when compared to many other Marines' careers.

Upon my return from Iraq, I was assigned to successive positions at The Basic School, Marine Corps University, and finally the U.S. Naval Academy that each included a teaching component. At each assignment, I was able to challenge future leaders to consider deeply their responsibilities and the effects they would have on their Marines' perspectives and understanding of the law of armed conflict and the rules of engagement. Given the opportunity to more deeply examine, reflect on, and regularly profess the importance of these two legal frameworks was a cathartic experience. As I engaged and challenged junior servicemembers, as well as those Marines poised to take command, I was reminded that the execution of war requires not only study but critical thinking, reflection, and ultimately the leadership of commanders. These assignments poignantly reminded me that, while experience may be the best teacher, study and reflection are necessary in transforming mere experience into knowledge and understanding. I was graced with the ability to share my experiences with thousands of Marines and translate those experiences into knowledge and understanding, not only for myself but for my fellow Marines.

Beyond considering others' experiences in combat, as with any endeavour, the Marine Corps and war are about people. As I reached the end of my career, I came to a personal realization that I have heard echoed by many career Marines. The sentiment is that "I don't always love the Marine Corps, but I always love my fellow Marines." I believe that to be true. Marines that read these stories will undoubtedly remember fellow Marines who they have loved and served with over the years. More importantly, I hope the American public will take this opportunity to learn about the extraordinary lives of ordinary Marines. In the case of Douglas A. Zembiec, readers will simply learn more about an extraordinary man. Doug's story is personal to me in that we attended the Naval Academy together and remained friends in the same circles during our Marine Corps careers. I did not know Doug intimately but well enough to realize his greatness, and I was fortunate enough to have a picture taken of him holding my firstborn son, a picture that I cherish to this day. It is men and women like Doug whom I got to know personally over the years and who have challenged me to be a better Marine, husband, father, friend, and even a better American. I aspire to be as great a man as Doug. It is the remembrance of Doug and the other contributors to this book who I anticipate will humble and inspire the American people, just as I am humbled and inspired when re-reading their stories.

For these reasons, I am proud that the Marine Corps University Press has provided this platform for these Marines and sailor to share their voices. This is an opportunity for our fellow Americans to learn about the more common experiences of war and the people who risked and sacrificed for our country. The collection strives to be self-reflective and, at times, even critical of the Corps. This was on purpose, and the editors of this book asked the writers to consider how they or how the Marine Corps could have better prepared or aided them to succeed. Critically evaluating your past experiences is part of being a modern Marine. Clearly, the contributors have a profound affection for the Corps, and the few points of critique included in their essays are given in constructive fashion. As Marines, we only want the Marine Corps to be better. On the few days we do not fully love the Marine Corps, we never stop loving our fellow Marines. I am certain that after reading this book, our fellow Americans will not only love the Marine Corps more deeply but understand and love the most extraordinarily ordinary Marines, who have volunteered to serve our nation.

Lieutenant Colonel Carroll J. Connelley (Ret)*

---

* LtCol Connelley was an infantry officer turned judge advocate who served in Iraq and Africa as an operational lawyer. He taught leadership, ethics, and law at The Basic School, Marine Corps University, and the U.S. Naval Academy. He is currently a department counsel at the Defense Office of Hearings and Appeals prosecuting security clearance cases.

PREFACE

The attacks on 11 September 2001 defined a generation of U.S. Marines. There are those who, that day, were already wearing the uniform, and there are those who felt called to join the Corps following the attacks. All carry with them the vivid memory of where they were that day and what that day meant for our nation. Over the course of more than a decade, thousands of Marines have deployed to Afghanistan, Iraq, and other parts of the world where they have fought against a persistent enemy. Over these years, many Marines cataloged their thoughts and observations in various ways, and their stories provide a much needed understanding on the impact combat has on those who experience it—even on those who are trained for it.

As a public affairs officer, I look for stories, and together with Paolo G. Tripodi, we looked for Marines to share theirs. Thus, we invited Marines and a U.S. Navy chaplain, who deployed with Marines, to write about their experiences of going to war. We gave them guidance about the structure for their chapters, but not about the content. What they chose to share and how they chose to express themselves reveals the diversity of events that stick with and follow veterans into stateside military service and civilian life.

This project particularly interested me because I too am a veteran of Iraq and Afghanistan. When someone asks me about my deployment experiences, I think of the color green; a green so rich and so bright that it hurts the eyes. When I left for Iraq in 2003, I embarked on a ship. The ship was gray. The sea was blue. The sky was blue. Iraq was tan—the sand, the buildings, sometimes the air too. We wore tan uniforms. When I sailed home, I was again surrounded by blue and gray. I had not seen the color green in months. When we were off the coast of North Carolina and the land was a thick dark line in the distance, I embarked on a gray helicopter into a blue sky to touch down on a grassy landing zone at Marine Corps Base Camp Lejeune, North Carolina. The back of the helicopter opened up, and I stepped out onto green grass surrounded by green trees in the thick heat of June. I was overwhelmed by color. I barely remember people welcoming me home because I could not stop looking, and yet could not look fully at the grass and trees. There was no deeper meaning in this memory. During my deployment, green had no particular meaning to me, and I had not noticed its absence until it was again a part of my environment. It was the startling aspect of noticing its reappearance in my world that struck me.

When people ask about my time at war, I do not share this story because I feel it would disappoint them. They do not want to hear about colors even though that is what comes to my mind first. I served in combat zones but not in combat—that is a distinction. I did not lead Marines through a field of gunfire; I did not clear houses of enemy combatants. I did not call for fire on the enemy from air, artillery, or ships. I never had to hold dying Marines in my arms and pray for them to survive while not showing them my fear that they would not make it home. I did see the bodies of people killed in combat and was struck by their stillness. I saw the wreckage of vehicles and had looping thoughts of what it must have been like to be in the vehicle when it was destroyed. I did see photos of carnage and then walked the ground where it occurred just days before. I felt heartache over hearing about the deaths of Marines in real time and felt sick about the bad news their families and friends would soon hear. I thought about the Marines back home who would have to deliver that news. These experiences do not strike me as special or even unique, certainly not worthy of sharing. They strike me as less than worthy to share because there are Marines, sailors, soldiers, and airmen who experienced much worse. I do not feel claiming empathy is much of an accomplishment.

The stories I do share tend to be about the absurd moments. In Kuwait, I remember being handed a cage of pigeons to keep with us in the back of a large truck and being told, "If the pigeons die, put your gas masks on." I remember initially there being a problem with some of the pigeons staging cage death matches until we found a Marine in the unit who had raised pigeons in his childhood. He could tell the difference between male and female pigeons, which was important for properly caging the birds. This Marine sat in a sea of pigeon cages, meticulously flipping over pigeons in order to segregate them. He became the official "pigeon sexer" of Task Force Tarawa.[1] I learned, in that moment as well as others, that Marines have many and varying talents. Later in Iraq, I remember lecturing my corporal that he was not allowed to sell me. When we walked into villages, he was approached several times with offers to trade me for livestock. He joked with me that if the price was right, he would consider it.

Just after Baghdad fell in 2003, there was a lull in the fighting in our region. We were in al-Kut, Iraq.[2] The Marines had discovered an old British World War I cemetery that had become neglected and overgrown. They decided to fix that. Marines repositioned headstones, cleared weeds, and picked up trash, and the Seabees constructed a large and ornate cross to place in the center of the small square.[3] We invited members of the British forces to come to the rededication ceremony. The cemetery was surrounded on three sides by buildings. Iraqis were on the roofs, were in the street, and were leaning out of windows to watch what we were doing. I stood on a small wooden dais with the other observers as the rededication occurred. Nice words were said about those who had gone before us. The British

---

[1] The 2d Marine Expeditionary Brigade was named Task Force Tarawa during the United States' 2003 invasion of Iraq.

[2] Al-Kut is in eastern Iraq along the Tigris River.

[3] Seabees is a Navy construction unit, an abbreviation for construction battalion (CB).

had sent a religious man, who wore the robes and tall hat of some denomination. He led us in prayer. The Iraqis quietly watched the proceedings; some even bowed their heads with us. When the prayer was over and we said, "Amen," the word was echoed by the Iraqis on the three sides. After a very brief pause, the Iraqis then broke into song. I think it must have been the only song they all knew in English. But there on that sunny day, the Iraqis serenaded the ceremony with a solemn singing of "Happy Birthday." It began quietly and, then, was sung with more gusto until the triumphant end when they all cheered and clapped. In Afghanistan, I remember similar things that struck me as absurd, or just funny. I remember pilots returning from sorties and being tickled that, when they flew in support of Italians, the Italians ended their radio communication with "happy so far?" The Marines would reply, "Copy, (call sign) is happy." One day, as they shared another story about "happy," someone asked, "Are you sure they're not saying, 'copy so far'?" The looks on their faces, when they considered that, were priceless. We imagined the Italians somewhere discussing how strange Marines were for their constant reporting of their state of happiness.

For me, the war was on two separate planes; I was both an actor and an observer. I, like the authors and others who have gone before, experienced a wide range of emotions while deployed, including the fear of the unknown (and the known), the dismay at the losses, the joy of camaraderie, the boredom of waiting, the monotony of meals, ready to eat, the pride in the actions of my Marines, the incredible nature of some of the situations I found myself in, and the state of being uncomfortable, dirty, exhausted, and sick. I feel guilty, at times, that my stories are silly and can seem to make light of a time that had a devastating effect on others. But I remember and share the stories above because they make me smile. For me, they do not take away the twinge I feel when I remember those who did not come back whole or those who did not come back at all. This volume is dedicated to those Marines.

In the introduction, Dr. Paolo G. Tripodi gives a brief history of the telling of war stories first from military historian, war correspondents then to the firsthand accounts from American servicemembers. Citing past conflicts of World War II and Vietnam, Tripodi notes the effect that public support for a war has on its combat veterans upon their return home from warzones. That hometown attitude coupled with the raw experiences of combat are what veterans wrestle with every day, Tripodi argues. For this reason, understanding wartime experiences, although important, is not easy for civilians.

In chapter 1, "Captain Doug Zembiec," Sergeant Major William S. Skiles shares a story from his time as a first sergeant for Echo Company, 2d Battalion, 1st Marine Regiment (1st Marines). As the chapter title reveals, the hero of his story is Captain Douglas A. Zembiec, then company commander. Zembiec was a Marine officer who distinguished himself in many areas, and excelled in leadership. The setting is a tough deployment to Iraq in 2004. For Skiles, this essay was a labor of love. "Humility" and "compassion" are the words most often used to describe Zembiec, "the lion of Fallujah," with ample evidence to support their appropriateness. In Fallujah, Zembiec and Skiles were a cohesive command team that led their Marines through intense fighting. The captain was larger than life but not larger than death. Skiles brings the reader through his first meeting with Zembiec, their deployment, and to the day of the notification of Zembiec's death.

Chapter 2, "The Unspoken Leadership Challenges of Command," written by Colonel Brian S. Christmas, focuses on the death of Marines and the impact on their units and families. Christmas's story is unique, centering not on the battlefield but on family—his own and the families of his Marines. Marine families are actual characters in his essay, not an afterthought, and are central to the mission, not bystanders. Christmas begins thinking of families and their roles before the 3d Battalion, 6th Marines' 2010 deployment to Marjah, Afghanistan, and includes casualty notifications in his staff's training. Families were never far from his thoughts. When deployed and faced with mounting casualties, Christmas shared his notes to the families of deceased Marines. Christmas does not recount tales of battle and bravery but focuses on their effects.

Chapter 3 also focuses on family with a very personal story shared by Captain Matthew C. Fallon. In "A Fellow Marine in Combat," Fallon describes his combat experience in Afghanistan with intense detail and compassion. The true focus of his story falls after the battle. Marines have described family and home as being put on pause when they deploy. For families who lose a loved one, that pause is everlasting. Fallon's story shares both sides, of going to war and of watching a loved one go to war. His younger brother followed him into the Marine Corps and became an infantry officer. In the essay, Fallon watches his brother board a bus for the airport amid a sea of Marines' family members. Fallon then deals with sporadic communication, and—just hours before Fallon's own deployment—he gets the phone call no one wants to receive, his brother was badly wounded. When he returns home, all that he left behind—including emotions—is pushed abruptly to the foreground.

In chapter 4, "The Cobra, the Convoy, and a Crisis of Faith," Lieutenant Colonel Wayne R. Beyer Jr., a pilot, reflects on his own conflicting thoughts and feelings regarding his role in the Iraq War. His story includes a scattering of meaningful moments—a moment with a priest from his childhood, a moment with a Marine fuel truck driver, and a moment with a fellow pilot. As a man of faith, Beyer finds that participating in war from the sky does not lessen the human impact. He focuses on his faith both in the beginning as he contemplates joining the Service and in the end as he comes to grips with what he did during the war. In the moment of battle, rote memorization and training are what carry him through.

In chapter 5, "The Bastards' Shepherd," Navy Commander Brian D. Weigelt, a chaplain with the 2d Battalion, 4th Marines, in 2004, tells his own story of faith and doubt. Weigelt went to the Iraq War unarmed, though his own personal safety was seldom at the top of his mind. Weigelt's time deployed to combat is fraught with self-doubt and concern that he would not be enough for his flock. The chaplain's tale is about sadness and strength, being strong for others while experiencing the same challenges. His story is about pride and humility, being proud of those with whom he served and feeling humble at being accepted by them.

In chapter 6, "Trust Among . . . Allies?," Staff Sergeant Michael R. Moyer provides an enlisted Marine's perspective on the war in Afghanistan. He writes an emotional story about his experience in 2004 working with the Afghan National Army and Afghan National Police. His story includes a numbness in recalling the hard and cold facts of the deployment. He begins with anticipation as his unit, the 1st Battalion, 2d Marines, trains. Once in country, he meets Afghans and realizes that his own face is just another in a long line that the Afghans have and will meet. Moyer writes of fear, not for himself but for the Marines in his charge. He shares an intense desire to go into battle—to test oneself—combined with an intense desire not to lose anyone. He describes children being blown up. He worries about creating civilian casualties. When he returns home, Moyer wrestles with bitter feelings of pride for doing what few have done and anger for others who have not.

In chapter 7, "A Woman in Charge," Major Aniela K. Szymanski offers a wartime story on a different side of the combat operations spectrum—what comes after the combat, missions that are enabled by the clearing out of enemy targets. Szymanski and her team deliver on the promises Staff Sergeant Moyer felt were not kept. Her story also gives insight into three things she contended with during her deployment: being a woman, being a reservist, and being in civil affairs. Ironically, she notes that the only people who took issue with her gender were senior Marines in positions of leadership. The Marines in her charge, other troops, and Afghan citizens were not concerned at all by her gender or felt more comfortable coming to her to share their observations, concerns, and needs. Her civil affairs mission was the true challenge as she encountered policies and Marines who gave the mission lip service, but failed to be attentive to or address the true requirements of success. She tells of farm projects that needed simple solutions that were difficult to attain because of bureaucracy and infantry Marines who were concerned with security. Her frustration is evident, but so too is her pride. Szymanski is proud of the Marines who set the conditions for her mission. She is proud of her team members and their ability to solve problems despite the constraints of the situation.

In chapter 8, "Crowd Control," First Sergeant Nicanor A. Galvan also shares Marines' frustrations, but is self-directed. Galvan looks back on his 2003 tour in Iraq, when he was a sergeant, with the patience and maturity of an older man. He gives a detailed story of one mission, which he had not trained to accomplish. The combat mission he trained for—the march to Baghdad, Iraq, the mission that put him where he was—serves as the prequel. When his squad is assigned the noncombat task of controlling a crowd outside of a bank, Galvan and his squad are challenged by a crowd of Iraqis who just want to be paid. He matter-of-factly shares his frustrations and, in reflection, can see that he did not handle the situation as well as he could have. His story is one of regret. He regrets how he treated the Iraqi people. The word training is bandied about in the military. His story shows the consequences of what can happen when the training does not match the mission.

Noncombat missions are not uncommon in combat zones, and some Marines are trained for them. As chapter 9's title reveals, "The Advisor" tells one such story. Colonel Daniel L. Yaroslaski uses the paving of a road to detail the nuances of working with Afghans and to explain how a comment he heard during training served him well in his mission. This rings true here. In Yaroslaski's story, he has plenty of exasperations with the way things are done. He admits mistakes and comes to appreciate what Afghans work with in terms of resources and expectations.

In chapter 10, "A Logistician's Preparation for Combat," Lieutenant Colonel Clifton B. Carpenter presents another noncombat mission. Carpenter, who served as the 24th Marine Expeditionary Unit (24th MEU) logistics officer during the unit's 2008 deployment to Afghanistan, focuses on three topics: the unit (he is an advocate of the MEU as a training ground); family (he is married to another Marine; in true logistician fashion, they war game their family readiness); and individual breaking points (he discovers true humility when he realizes he understands very little of what his subordinates are responsible for). I actually deployed with Carpenter. Any doubts he had about his ability were transparent. I, as a captain, thought then-Major Carpenter was one of the strongest and most capable officers on the staff. Nothing was too difficult for him to make happen. And if he thought something was going to be impossible or difficult, I never saw it reflected in his words or actions. I had no doubts the Marine expeditionary unit would get where it needed to be with what it needed.

Chapter 11, "A Perspective on Leadership Attributes in Combat," explores the responsibilities of Major Benjamin P. Wagner, who as a captain in Afghanistan had an incredible duty in terms of area, numbers, and missions. His essay focuses on leadership, and he tells the stories of fellow Marines who deployed and embodied the qualities he values—competency, grace, dignity, and judgment.

Lieutenant Colonel Kelly Frushour*

---

* LtCol Frushour is a public affairs officer (PAO) with the U.S. Marine Corps. She has served as the PAO for 3d Marine Division, 2d Marine Division, 2d Marine Expeditionary Brigade (during its Iraq deployment in 2003), 24th Marine Expeditionary Unit (during the unit's Afghanistan deployment in 2008), and II Marine Expeditionary Force. She has also served as an aide-de-camp for the deputy commander of Fleet Marine Forces Atlantic, Marine Forces South, and Marines Forces Europe; as the speechwriter for the Assistant Commandant of the Marine Corps; as an instructor/company commander at The Basic School; and, at that time of this publication, as the commanding officer of Region 7 (North and West Africa), Marine Corps Embassy Security Group.

ACKNOWLEDGMENTS

This acknowledgment should start with a special thanks to my coeditor, Kelly; to a dear friend, Carroll, who has always been ready and happy to discuss this and other projects with me; and to all the authors for their contributions. If this was a normal book, it would just be appropriate to thank them for their commitment to preparing this book, but this is not the culmination of a normal project. The authors, who have provided so many personal, intimate, and enlightening contributions, are Marines and a Navy chaplain who have served in war. We acknowledge and thank them, first and foremost, for their service, their decision to sacrifice their lives, if necessary, and the hardship they put their families through when they decided to serve, especially when deployed in harm's way. For what they experienced and went through—thank you! We are indebted to them all for their decision to write about their experiences going to war, not an easy thing to do. They made this book possible.

I am particularly grateful to Marine Corps University Press (MCUP). What a great crew. MCUP is a young, yet quickly maturing, professional press. Angela Anderson, the senior editor, Alexandra Kindell, the acquisitions editor, and Jennifer Clampet, the manuscript editor, through their commitment, passion, and time supported this project. They have been instrumental in making the book such a solid product. Graphic designer Robert Kocher's creativity continues to amaze and impress me.

Both Colonel Scott Erdelatz and Dr. James Van Zummeren, the Lejeune Leadership Institute director and deputy director, have always encouraged and supported this project. Their constant interest in the book has been extremely valuable. They have provided, and continue to provide, the type of leadership that allows projects like this flourish.

Finally, I owe a special thanks to my wife, Jenny, and my son, Antonio, for their love and constant support.

Dr. Paolo G. Tripodi

LIST OF ABBREVIATIONS

AAR	After action report
AAV	Amphibious assault vehicle
ANA	Afghan National Army
ANP	Afghan National Police
ANSF	Afghan National Security Force
AO	Area of operation
AOR	Area of responsibility
APOBS	Antipersonnel obstacle breaching system
AWS	Amphibious Warfare School
BAS	Battalion aid station
BDA	Battle damage assessment
BLT	Battalion landing team

CAAT	Combined antiarmor team
CAG	Civil affairs group
CAS	Close air support
CAX	Combined arms exercise
CFT	Combat fitness test
COC	Combat operations center
COIN	Counterinsurgency
COP	Combat outpost
DASC	Direct air support center
DASC(A)	Direct air support center (airborne)
DEA	U.S. Drug Enforcement Administration
DM	Designated marksman
DSF	District Stability Framework
ECM	Electronic countermeasure
ECP	Entry control point
ETT	Embedded training team
EWS	Expeditionary Warfare School
FAC	Forward air controller
FARP	Forward arming and refueling point
FOB	Forward operating base
FRO	Family readiness officer

GWOT	Global War on Terrorism
H&S	Headquarters and support (company)
HET	Human intelligence exploitation team
HMLA	Marine light attack helicopter squadron
HMLAT	Marine light attack helicopter training squadron
HMMWV	High Mobility Multipurpose Wheeled Vehicle or humvee
IED	Improvised explosive device
IOC	Infantry Officer Course
ISAF	International Security Assistance Force
ISCI	Interim Security for Critical Infrastructure
ISR	Intelligence, surveillance, and reconnaissance
LAR	Light armored reconnaissance
LAV	Light armored vehicle
LOE	Line of effort
LOI	Letter of instruction
LOO	Line of operation
LP	Listening post
LZ	Landing zone
MAGTF	Marine air-ground task force
MarDiv	Marine division
MARSOC	Marine Corps Special Operations Command

MATV	MRAP all-terrain vehicle
MCAS	Marine Corps air station
MCB	Marine Corps base
MCCS	Marine Corps Community Services
MCDP	Marine Corps Doctrinal Publication
MCWP	Marine Corps Warfighting Publication
MEF	Marine expeditionary force
MEU	Marine expeditionary unit
MEU (SOC)	Marine expeditionary unit (special operations capable)
MOPP	Mission-oriented protective posture
MOS	Military occupational specialty
MRAP	Mine-resistant ambush protected vehicle
MRE	Meals, ready to eat
MSR	Main supply route
MWSS	Marine wing support squadron
NATO	North Atlantic Treaty Organization
NBC	Nuclear, biological, and chemical
NCO	Noncommissioned officer
NEO	Noncombatant evacuation operation
NROTC	Naval Reserve Officer Training Corps
NVG	Night vision goggles

OCS	Officer Candidate School
OEF	Operation Enduring Freedom
OIF	Operation Iraqi Freedom
OP	Observation post
OSO	Officer selection officer
PAO	Public affairs officer
PB	Patrol base
PCS	Permanent change of station
PFC	Private first class
PFT	Physical fitness test
PID	Positive identification
PITMOV	Preparation is the mother of victory
PLC	Platoon Leaders Class
PME	Professional military education
PRT	Provincial reconstruction team
PSD	Protective security detachment
PTSD	Post-traumatic stress disorder
QRF	Quick reaction force
RC	Regional Command
RCT	Regimental combat team
RPG	Rocket-propelled grenades

RTB	Return to base
Seabees	Navy construction unit (from CB for construction battalion)
SNCO	Staff noncommissioned officer
SOC	Special operations capable
SOTG	Special operations training group
TBI	Traumatic brain injury
TBS	The Basic School
UAV	Unmanned aerial vehicle
UDP	Unit deployment program
UPFRP	Unit, Personal, and Family Readiness Program
USAID	U.S. Agency for International Development
WTB	Warrior transition briefs
WTI	Weapons and tactics instructor
XO	Executive officer

INTRODUCTION

THE EXPERIENCE OF GOING TO WAR:
WORLD WAR II, VIETNAM, AFGHANISTAN, AND IRAQ
PAOLO G. TRIPODI

The experience of going to war, whether in a combat zone or in a support role, is extremely powerful. It is an experience that defines all who have served. Servicemen and women who have been to war have gone through what can be considered a wide range of highly amplified emotions. Those among us who have never been at war struggle to understand how intimate and powerful such an experience can be and to grasp what veterans have gone through, what they did, saw, and felt. Soldiers and Marines experience contradictory and, sometimes, disturbing feelings. On the one hand, they may fear being killed, but also be elated or saddened at taking the life of another. As so many have written, those in battle oscillate between the ensuing adrenaline rush from the excitement of combat and the extended periods of boredom. Participating in war requires patience with the "hurry up and wait" mentality as well as the unique type of boredom that emerges even in the midst of danger. Tim O'Brien, a Vietnam veteran, explained it best in his novel *The Things They Carried*:

> Even in the deep bush, where you could die any number of ways, the war was nakedly and aggressively boring. But it was a strange boredom. It was boredom with a twist. The kind of boredom that caused stomach disorders. You'd be sitting at the top of a high hill . . . and the day would be calm and hot and utterly vacant, and you'd feel the boredom dripping inside you like a leaky faucet, except it was not water, it was a sort of acid, and with each little droplet you'd feel the stuff eating away at important organs. You'd try to relax . . . Well, you'd think, this isn't so bad. And right then you'd hear gunfire behind you and your nuts would fly up into your throat and you'd be squealing pig squeals. That kind of boredom.[1]

These are emotions that veterans of battle understand while most cannot even fathom the depths of despair and exhilaration felt in these moments.

The desire to understand the human experience of going to war has inspired the work of outstanding and gifted writers, their books have given us a grasp of that experience. Several authors have provided us with a profound understanding of how war impacts human beings. John Keegan, one of the most distinguished and prolific British military historians, opens his book, *The Face of Battle: A Study of Agincourt, Waterloo, and the Somme*, with an enlightening narrative. Keegan begins his exploration of the experience of battle by emphasizing the difficulty of understanding such experience for those who have not lived it directly. "I have not been in a battle; not near one, nor heard one from afar, nor seen the aftermath," he wrote. Yet Keegan studied that experience with rare intellectual passion. "I have read about battles . . . have talked about battles, have been lectured about battles, and. . . . have watched battles in progress, or apparently in progress on the television screen." Not only did Keegan study military history and battles for nearly two decades as a lecturer at the British Army Royal Military Academy-Sandhurst, he was also instrumental in the education of several

---

[1] Tim O'Brien, *The Things They Carried* (New York: Broadway Books, 1998), 34.

generations of British Army officers. He had lived among those who had direct experience with war, combat, and battles, and yet he felt the need to share with readers of his book that, "I grow increasingly convinced that I have very little idea of what a battle can be like."[2] However, Keegan found a way to describe battles from eras long gone with meaning and insight.

Other scholars have combined their military experience with their academic knowledge. In *Acts of War: The Behavior of Men in Battle,* Richard Holmes, another British historian, takes the reader through a fascinating and eye-opening journey, exploring areas that are emotionally similar for soldiers throughout many modern conflicts.[3] Unlike Keegan, Holmes is not concerned about his lack of wartime experience. Parallel to his academic achievements, Holmes had a military career, and although he never deployed to a war zone, he felt comfortable in exploring the impact of war on the individual soldier. At a young age, Holmes joined the British Territorial Army and eventually became a brigadier general. Holmes provides an insightful analysis of the experience of war, with its complexities, including psychological, and consequences for soldiers of all backgrounds. In the 1980s, Keegan and Holmes teamed up to produce the well-received 13 episodes of the British Broadcasting Corporation's documentary *Soldiers: A History of Men in Battle.* Indeed, Keegan's and Holmes's books and their additional television work provide an excellent understanding, at least in part, of the war experience.

Until the twenty-first century, it was mainly historians, specifically military historians, who explored, analyzed, and wrote about the experiences of war. In smaller but nonetheless important numbers, war correspondents and journalists wrote about that war experience. One of the most respected war reporters was Ernest T. "Ernie" Pyle, whose articles and chronicles of World War II appeared in several newspapers. Pyle's columns from June 1943 to September 1944 were published in the book *Brave Men,* which tells

---

[2] John Keegan, *The Face of Battle: A Study of Agincourt, Waterloo, and the Somme* (New York: Penguin Books, 1978), 13.
[3] Richard Holmes, *Acts of War: The Behavior of Men in Battle* (New York: Simon & Schuster, 1985).

the story of soldiers at war. Pyle lived firsthand the experience of combat in Africa, Europe, and later the Pacific. While on this final assignment, Pyle who was then attached to the U.S. Army 77th Infantry Division on an island close to Okinawa, Japan, was killed by Japanese fire. He was posthumously awarded the Purple Heart. In *Brave Men*, Pyle made no effort to disguise his respect and sympathy for the infantry soldiers more than members of any other Service. Pyle stressed how the frontline soldier

> lived for months like an animal, and was a veteran in the cruel, fierce world of death. Everything was abnormal and unstable in his life. He was filthy dirty, ate if and when, slept on hard ground without cover. His clothes were greasy and lived in a constant haze of dust, pestered by flies, and heat, moving constantly, deprived of all things that once meant stability—things such as walls, chairs, floors, windows, faucets, shelves, Coca-Colas, and the little matter of knowing that he would go to bed at night in the same place he had left in the morning.[4]

Pyle wrote of the soldiers' lives as if their grit infected the ink in his pen.

Indeed, those who deployed during World War II, in addition to facing horrifying battlefield situations, often lived in extremely uncomfortable conditions. Many GIs who landed in North Africa in 1943—and were lucky to survive the terrible battles—fought for a few years in Italy, France, and Germany until the war was over.

Much like those GIs who faced the horrible challenges of landing in Normandy on D-Day, young Marines in the Pacific theater were in a similar situation. They faced the horror of multiple landings in places fiercely and heavily defended by Japanese soldiers who were fully aware that any inch of ground given up brought the United States closer to victory.

Unlike historians, and in a different way from war correspondents, these soldiers and Marines developed a much more emotional and personal understanding of such a powerful experience. Those veterans who decided to share their experience were in a unique position to provide the story from inside. They had experienced death, killing, fear, grief. They walked

---

[4] Ernie Pyle, *Brave Men* (New York: Henry Holt and Co., 1944), 5.

alongside evil and, in some cases, stepped right into it. For many of them, the intensity of the emotions and circumstances were extremely traumatic. Several of the most insightful books by World War II veterans were written and published decades after the end of the conflict. Although time never fully healed the impact the war made on them, at least it made them able to cope with their past and reflect on their experience of war. With time, they were able to share such an experience with those who never had a taste of it.

J. Glenn Gray of Mifflintown, Pennsylvania, was one of the many GIs for whom service in the U.S. Army started a few months before the Japanese attack on Pearl Harbor in December 1941 and ended months after Berlin had fallen in May 1945. During those four years, Grey participated in several campaigns and battles and experienced a range of powerful emotions. His book, *The Warriors: Reflections on Men in Battle*, was published in 1954, nearly a decade after the end of the war. Gray was a particularly qualified observer of men at war, not only because of his extensive experience, but also because of his educational background. He had received his PhD in philosophy from Columbia University, while he was also drafted in the U.S. Army, and after the war, he became a philosopher and a professor at Colorado College. His observation of men at war is compelling and continues to enlighten readers today. Gray provides an intimate analysis of how conflicting emotions and highly unusual behavior became normal for soldiers who faced the challenges of landing on enemy territory, were targeted by enemy fire, and lived through enemy air raids. In anticipation of landing, soldiers' "[m]oods of fear, anticipation, helplessness, praying and cursing, adventure and longing succeed each other like lightning."[5] When describing the impact of combat on the individual soldier, Gray notes how, in the same individual, inhuman cruelty could easily be replaced by superhuman kindness. He wrote that "Again and again in moments of this kind I was as much inspired by the nobility of some of my fellows as appalled by the animality of others, or, more exactly, by both qualities in the same person."[6] Gray stresses how the experience of war and combat removed many, if not all, inhibitions from those

---

[5] J. Glenn Gray, *The Warriors: Reflections on Men in Battle* (Lincoln, NE: Bison Books, 1998), 13.
[6] Ibid., 14–15.

men. Their unfiltered behavior was spontaneous, a reaction to the highly emotional environment. Those millions of men who learned to live in such a highly charged and often confusing emotional environment, "have discovered in it a powerful fascination . . . many men both hate and love combat. They know why they hate it; it is harder to know and to be articulate about why they love it."[7]

While Gray was 30 years old when he arrived in North Africa in 1942, Eugene Sledge of Mobile, Alabama, was not yet 18 when he joined the U.S. Marine Corps. Sledge dropped out of an officers training program and enlisted to ensure he would deploy and join combat against the Japanese. "Sledgehammer," as he was called by his fellow Marines, participated and survived two bloody landings in which thousands of Marines were killed: Peleliu in the Palau Islands and Okinawa. His experiences were terrifying and likely also traumatic. Against regulations, while deployed in combat, he recorded those experiences in notes scribbled in a pocket-size copy of the New Testament that he always carried with him. In 1981, more than three decades after Sledge left active duty, he published *With the Old Breed: At Peleliu and Okinawa*. The book's brief preface provides an excellent key to understanding Sledge's mind-set as he worked through his memories and recollections of the war.

> My Pacific war experiences have haunted me and it has been a burden to retain this story. But time heals, and the nightmares no longer wake me in a cold sweat with pounding heart and racing pulse. Now I can write this story, painful though it is to do so. In writing it I am fulfilling an obligation I have long felt to my comrades in the 1st Marine Division, all of whom suffered so much for our country. None came out unscathed. Many gave their lives, many their health, and some their sanity. All who survived will long remember the horror they would rather forget.[8]

---

[7] Ibid., 28.

[8] E. B. Sledge, *With the Old Breed: At Peleliu and Okinawa* (New York: Ballantine Books, 2010), xiii–xiv.

The Pacific theater was terrifying. The anticipation for Sledge's first combat experience, the landing on Peleliu, was overwhelming. With his fellow Marines ready to land on the small island, he was ordered to get on the landing vehicle and wait for a Navy officer to give the order to head toward the beach.

> We waited a seeming eternity for the signal to start toward the beach. The suspense was almost more than I could bear . . . I broke out in a cold sweat as the tension mounted with the intensity of the bombardment. My stomach was tied in knots. I had a lump in my throat and swallowed only with great difficulty. My knees nearly buckled. I clung weakly to the side of the tractor. I felt nauseated and feared that my bladder would surely empty itself and reveal me to be the coward I was. But the men around me looked just about the way I felt.[9]

Finally, as Sledge was going through all these emotions, the Navy officer gave the signal that they could begin moving toward the beach. The noise of the amphibious tractor engine became extremely loud and the Marines went toward a "frightful spectacle," seemingly drawing them "into the vortex of a flaming abyss."[10]

For those young Marines who were lucky to survive the "flaming abyss," life would never be the same. The fear of being killed was matched only by the fear of losing a fellow soldier or Marine. The bond among fellow soldiers created by war has been and always will be very strong. In such an environment, individuals are ready to do unthinkable things and deeds for each other.

The young Americans who volunteered, or were drafted, to fight in World War II were motivated by the Japanese attack on Pearl Harbor. The unprovoked attack against the U.S. naval base in the Pacific was a motivating and inspiring factor for the American public overall. Yet, the attack only convinced Americans why the nation should face the Japanese in battle, but that conviction was not immediately connected with the war that raged in

---

[9] Ibid., 55–56.
[10] Ibid., 55.

Europe at the same time. As information became known about Nazi brutality against the Jewish people and civilians in the occupied countries, the war increasingly became portrayed as one of "good" versus "evil." Many young Americans wanted to be a part of it. General Dwight D. Eisenhower, more effectively than many of his contemporaries, captured the sentiment of a majority of servicemembers who fought in the European theater. After visiting many concentration camps and witnessing the magnitude of the evil perpetrated in those terrible places, he said, "We are told that the American soldier does not know what he is fighting for. Now, at least, he will know what he is fighting against."[11] Eisenhower's feeling was reflected and shared by soldiers on the ground. In April 1945, after liberating Kaufering concentration camp, Army Major Richard D. "Dick" Winters of New Holland, Pennsylvania, and his paratroopers of the 101st Airborne Division were overwhelmed by the horror.[12] The images of death and immense suffering were difficult to comprehend. The now unveiled project of genocide provided instant clarity on why those American GIs had fought and died for months against the Nazis. Winters wrote:

> As I went through the war it was natural to ask myself, *Why am I here?* *Why am I putting up with the freezing cold, the constant rain, and the loss of many comrades? Does anybody care?* A soldier faces death on a daily basis and his life is one of misery and deprivation. He is cold, he suffers from hunger, frequently bordering on starvation. The impact of seeing those people behind that fence left me saying, if only to myself, *Now I know why I am here! For the first time I understand what this war is all about.*[13]

American servicemembers in Europe and in the Pacific, were joyously welcomed by the population of the countries that they liberated from the brutal occupation of the Germans and Japanese. With their jobs done, they were celebrated upon their return to the United States. Newspaper articles, movie

[11] Gen Eisenhower as quoted in Michael S. Sherry, *In the Shadow of War: The United States Since the 1930s* (New Haven, CT: Yale University Press, 1995), 91.

[12] Kaufering and Mühldorf were two subcamps of the Dachau concentration camp in Bavaria.

[13] Maj Dick Winters, *Beyond Band of Brothers: The War Memoirs of Major Dick Winters*, contrib. Col Cole C. Kingseed (New York: Berkley Caliber, 2006), 215. Emphasis in original.

reels, and parades announced the success of the Allies and its American members. Soldiers and Marines enjoyed a sense of great accomplishment, and society did not miss an opportunity to acknowledge what they had done, and their sacrifices, missing no occasion to celebrate them.

A few decades later, the American society struggled to understand the U.S. commitment in Vietnam and so did many of the young Americans who were sent to fight there. During the Vietnam war, servicemembers' experiences differed significantly from those of World War II veterans. Two sociologists, Ellen Frey-Wouters and Robert S. Laufer, investigated how the generation that went to Vietnam perceived war experiences. In *Legacy of a War: The American Soldier in Vietnam,* they note that "Vietnam was a strange and dangerous place for American soldiers" and "veterans admitted that upon arrival in Vietnam they experienced doubts about why they were there. Questions arose in their mind about the meaning of the war. They felt they were 'the liberators,' but [unlike the World War II veterans] they have no clear picture of what 'the liberation' was all about."[14] For many of the interviewed veterans, their first impressions stemmed from a negative reaction of the locals despite the servicemembers altruistic goals.[15]

O'Brien of Austin, Minnesota—a Vietnam veteran who served with the U.S. Army—was among those who had a hard time understanding the U.S. military commitment in that region of the world. Shortly after graduating from Macalester College, he was drafted to fight a war he "hated." In his mind, "[T]he American war in Vietnam seemed to me wrong. Certain blood was being shed for uncertain reason . . . The very facts were shrouded in uncertainty: Was it a civil war? A war of national liberation or simple aggression? Who started it and when and why?"[16] O'Brien's sentiment about the war was not unique. The impact the war had on American society was well described by another veteran, former Marine Lieutenant Philip Caputo of Westchester, Illinois. In Caputo's view, "Vietnam was the epicenter of a

---

[14] Ellen Frey-Wouters and Robert S. Laufer, *Legacy of a War: The American Soldier in Vietnam* (London: Routledge, 1988), 5.

[15] Ibid., 6.

[16] O'Brien, *The Things They Carried*, 40.

cultural, social, and political quake that sundered us like no other event since the Civil War."[17]

For American soldiers and Marines, the battlefields in Vietnam were significantly more chaotic than the ones veterans faced during World War II, and certainly the Vietnam War was more complex and ambiguous. Soldiers and Marines deployed to the Southeast Asian country encountered volatile frontlines. Often there were none. The enemy they faced was a combination of regular and guerrilla forces. Frequently, combat took place in an environment with a significant presence of civilians, who may have been neutral, but often were hostile to the American troops. Describing the conflict in Vietnam, Caputo wrote that it "combined the two most bitter forms of warfare, civil war and revolution, to which it was added the ferocity of jungle war."[18] Such an extremely hostile environment—a source of constant frustration and fear—quickly took a toll on the troops. Ghostly snipers, booby traps, and mines were common features of the war environment. Difficulty in identifying the enemy and hitting them was a major frustration. The enemy, who seamlessly blended with the civilian population, applied an irritating hit-and-run tactic. Anybody could be the enemy. Caputo stressed that "Some men could not withstand the stress of guerrilla fighting: the hair-trigger alertness constantly demanded of them, the feeling that the enemy was everywhere, the inability to distinguish civilians from combatants created emotional pressures which built to such a point that a trivial provocation could make these men explode with the blind destructiveness of a mortar shell."[19]

The cumbersome experience of fighting in the jungle wore on the troops' nerves and psyches. The enemy proved difficult to spot and to engage. For O'Brien, the enemy was a ghost that came out late at night when "it seemed that all of Vietnam was alive and shimmering—odd shapes swaying in the paddies, boogiemen in sandals, spirits dancing in old pagodas." To the American soldier, the enemy ghost "could blend with the land, changing

---

[17] Philip Caputo, *A Rumor of War* (New York: Macmillan, 1996), 353.

[18] Ibid., xviii.

[19] Ibid., xix.

form, becoming trees and grass. He could levitate. He could fly. He could pass through barbed wire and melt away like ice and creep up on you without sound of footsteps. He was scary."[20] O'Brien provides an elegant and, at the same time, accurate description of what the encounter with the enemy was like and how terrifying these experiences were. Employing guerrilla warfare, the astute and versatile enemy became invisible among the Vietnamese people and seemed to be gifted with magical powers used to attack and kill Americans. In a short period of time, thousands of young American troops were killed in Vietnam, and more feared the same fate.

It did not take long for this grim reality to be known to Americans back home.

The Vietnam War became the first televised conflict. New technology and the growing popularity of home television sets allowed Americans to view daily coverage of the conflict. Journalists and war correspondents used the advancing medium in an effective way, providing Americans with constant updates of the conflict and the images that often graphically narrated death and destruction. By 1960, nearly 90 percent of American homes had a television set, priming the country for a different kind of dialogue between society and the government about the conflict.[21] Opposition to the war that had started on a small scale in 1965 had grown significantly by 1967. In that year, more than 15,000 troops were killed and more than 100,000 were wounded. Veterans who returned home, however, encountered a divided society that provided, at best, a lukewarm welcome and, at worst, hostile confrontations.

Karl Marlantes of Astoria, Oregon—a Vietnam veteran Marine lieutenant and the author of *What It Is Like to Go to War* (the memoir that actually inspired the premise for this book)—shared his painful experience of coming back from Vietnam and being confronted by hostile people. While walking in uniform in Washington, DC, he was followed and insulted by a group of young Americans. At Union Station, he boarded a train to New

---

[20] O'Brien, *The Things They Carried*, 202.
[21] Katie McLaughlin, "5 Surprising Things that 1960s TV Changed," CNN News, 25 August 2014, http://www.cnn.com/2014/05/29/showbiz/tv/sixties-five-things-television/.

York. While seated during the trip, a woman proceeded to "get up and come down the aisle. She was looking right at me, lips pressed tight. She stood in front of me and she spit on me . . . I was trembling with shame and embarrassment . . . I wiped her spit off as best I could and pretend to go back to reading trying to control the shaking."[22] For many Vietnam veterans, such as Marlantes, the public hostility came at a time when servicemembers desperately needed their fellow Americans' support. Unlike the World War II veterans who returned to an elated society offering the warmest welcome, Vietnam veterans were ostracized. Their return was not marked with national celebrations as was given for those who fought the righteous war against fascism and genocide.

Nearly three decades later, a new generation of soldiers and Marines enjoyed an outcry of public support for the war the United States fought in Afghanistan and later on in Iraq. Like those who responded to the call to arms after Pearl Harbor, the infamous terrorist attacks of 11 September 2001 motivated young Americans to fight with spirit in one of the country's longest wars. In many ways, the experience of the terrorist attacks on 11 September, or 9/11, were similar to the attack against Pearl Harbor in 1941, but amplified by a few significant differences. The 9/11 attacks targeted mainly civilian facilities with symbolic values, with the exception of the Pentagon. While unacceptable, the attack against the U.S. Navy Pacific Fleet in 1941 made strategic sense as it intended to debilitate the more powerful nation while Japan solidified its position in Asia. In contrast, the 9/11 attack targeted harmless civilians and caused the most death and devastation the country had ever seen on home soil. The event was also caught on camera. Many Americans who had been alerted by the plane hitting the first tower turned to their televisions and witnessed the second attack. It was clear that what happened on 9/11 was much more than an infamous terrorist attack; it was a declaration of war. American society was painfully wounded and, at the time, it received the sympathy and support of the world community.

---

[22] Karl Marlantes, *What It Is Like to Go to War* (New York: Atlantic Monthly Press, 2011), 177.

The U.S. military that began active preparation to go to war against al-Qaeda following the 9/11 attack was a significantly different organization from the one that had entered World War II and later on the Vietnam War.

The American military for decades had moved away from the draft to create an all-volunteer force, made up with individuals who committed themselves to serve. In addition, the number of women in uniform, as well as the new capacities in which they could serve, steadily increased. Thousands of women served in both Afghanistan and Iraq, a significant number participated in combat, hundreds were killed, and hundreds received the Combat Action Ribbon.

The 2001 attacks inspired thousands of young Americans to enlist, to become active players in punishing those responsible for the death and destruction of 9/11, and to prevent future attacks against Americans. The new military recruits had a sincere desire to serve the country in the most dangerous and most honorable way. In a matter of weeks, they were ready for war. In less than a month, the U.S. military began operations against the Taliban government in Afghanistan.

The "wars" they were about to fight would present them with old and new challenges. The initial belief of many senior political and military leaders—that the war would last for just a short period of time—would quickly prove to be wrong. For a relatively small, all-volunteer force, sustaining a protracted commitment in two major conflicts would prove to be a major challenge. As a result, after more than a decade at war, thousands of U.S. servicemembers deployed more than once. The multiple deployments took a significant toll on them and their families.

As troops began their deployments, a considerable number of journalists and war correspondents went along with U.S. and international forces, many of whom were "embedded." In Afghanistan and later on in Iraq, the number of journalists and war reporters grew dramatically in comparison to previous conflicts. The availability of new communication technologies made reporting from war zones a significantly easier job, even though the war environment was, in many ways, very dangerous to them. In addition to the daily threats posed by working in a war environment, terrorist organizations preyed on them as easy targets and had no hesitation about killing them. Sadly, the number of reporters killed in war zones indeed grew. The current generation of war correspondents, in many cases, wanted to provide a better and more intimate exploration of the psychological and emotional impact that the war has on the individuals who fight it. One of the most accomplished and celebrated war correspondents, Sebastian Junger, deployed several times and for months with Army units in the middle of combat zones in Afghanistan. Junger's reporting was enlightening because he was in the midst of the battles that raged from the roads to towns as well as more traditional battle sites. In such books as *War* and such documentaries as *Restrepo* and *Korengal*, Junger provides a powerful narrative of the complexities of being at war.[23]

While veteran reporters returned to inform the public, a new generation of veterans from various U.S. military branches also came home to reflect upon their experiences. These veterans provided the most compelling narratives about the wars, sharing their experiences with fellow servicemembers and the American public. In their books and essays, they explore the complexities and challenges of fighting a different and unusual war—counterinsurgency—from a firsthand perspective.

In Afghanistan and Iraq, soldiers and Marines operated in exacting environments. In the normal course of operations, they would witness harm and misery inflicted on women, children, and other defenseless civilians. Such experiences have proved to be extremely powerful. The inability to do much or anything at all to protect those in harm's way is highly frustrating. The enemy's use of suicide bombers, snipers, and improvised explosive devices (IEDs) was problematic and unpredictable and made protecting comrades and civilians complicated.

One of the most difficult challenges for the troops that served in both Afghanistan and Iraq was to be able to transition from what would be considered a traditional soldier's role to a number of roles that soldiers and Marines operating in a counterinsurgency environment must be able to perform. After main military operations were over, soldiers and Marines

---

[23] Sebastian Junger, *War* (New York: Twelve, 2011).

were tasked to provide humanitarian assistance, perform crowd control, back up law enforcement, and support and protect the local population, often an uncertain and sometime hostile population. In this process, they were still targets for terrorists, possibly easier targets as their roles shifted away from fighting to helping. The experience could be described mildly as frustrating, but in reality it was nerve wrecking. At any moment, at any intersection, one could be killed or brutally maimed, and thus the sight of any garbage bag on the side of the road, potentially an IED, could be the cause of great anxiety. The absence of a frontline made it so that once out of the "security of the base," an invisible enemy could strike, kill, and remain unpunished.

In 2004, civilian combat historian Patrick K. O'Donnell joined 1st Platoon, Lima Company, 3d Battalion, 1st Marines, in Iraq as the unit engaged in urban fighting during the second battle of Fallujah. O'Donnell describes the urban fighting as "extremely personnel-intensive and, in terms of casualties, one of the most expensive military operations. Clearing building is combat at its most primitive. The fight is up close and personal . . . While the Marines' training and technique improved the odds, clearing was inevitably reduced to a high stakes game of Russian roulette. Kick the door in and see what's inside: either it's empty or there's a machine gun pointed at your head."[24]

O'Donnell's description of urban fighting is echoed in *Redeployment* by Phil Klay of Westchester, New York, a former Marine captain who served in Iraq. He wrote that "In a city there's a million places they can kill you from. It freaks you out at first. But you go through like you were trained and it works."[25] Enemy forces in Iraq and Afghanistan found methods of killing or maiming Americans that brought a new type of deadliness and unpredictability to war.

Dealing with IEDs was probably the most difficult thing to do. Tyler E. Boudreau, a former Marine captain, describes the impact of the use of IEDs made on servicemembers as "the inflictor of more casualties than any other

---

[24] Patrick K. O'Donnell, *We Were One: Shoulder to Shoulder with the Marines Who Took Fallujah* (Cambridge, MA: Da Capo Press, 2006), 81.

[25] Phil Klay, *Redeployment* (New York: Penguin Books, 2014), 12.

kind of enemy attack, bar none . . . It was what every soldier would remember about the war, what every soldier feared most about the war. In our Area of Operations, every drive down every road was a round of Russian Roulette."[26] The comparison to Russian roulette is a clear analogy. Conducting patrols was a gamble. No matter how well trained and equipped, no matter the intent of the mission, whether on patrol to get the enemy or to help build a school or a hospital, servicemembers were constantly risking their lives.

After operating in these dangerous environments, the Afghanistan and Iraq generation of veterans has strung together narratives of frustration and fear along with more traditional stories of camaraderie and pride in their work. They feel a strong sense of brotherhood with their comrades, and believe they made a significant difference in the local population where they served. They are proud of what they accomplished and, in large numbers, American servicemembers say they would do it all again. Yet veterans have expressed pain over a perceived apathy or distance that the American public developed toward the wars. Veterans "have come back to a nation that has embraced them—warmly, strongly, positively—and put tremendous value and appreciation into their service," said former U.S. Secretary of Defense Charles T. "Chuck" Hagel.[27] However, many of these servicemembers did not feel the public embrace the war. According to a 2014 *Washington Post* and Kaiser Family Foundation poll, nearly "seven in 10 [veterans] feel that the average American routinely misunderstands their experience, and slightly more than four in 10 believe the expressions of appreciation showered upon veterans—often at airports, bars and sporting events—are just saying what people want to hear. More than 1.4 million vets feel disconnected from civilian life."[28]

The causes for the disconnect between veterans and those who attempted to welcome them home from the front is difficult to pinpoint, but America's historical memory about war can offer some answers. World War II provided

---

[26] Tyler E. Boudreau, *Packing Inferno: The Unmaking of a Marine* (Port Townsend, WA: Feral House, 2008), 45.

[27] Rajiv Chandrasekaren, "A Legacy of Pain and Pride," *Washington Post*, 29 March 2014, http://www.washingtonpost.com/sf/national/2014/03/29/a-legacy-of-pride-and-pain/.

[28] Ibid.

a benchmark for what Americans consider a "good war." Sixteen million men went off to war, leaving behind women, children, and the elderly. Almost every American civilian knew someone who went off to war. The American public was invested in the war effort with rationing, working in factories, collecting recyclables, and buying war bonds. At a time when the nation was coming out from the painful 10-year long Great Depression, "victory" over tangible and scary foes permeated American culture. Victory was attained and marked by VJ Day (or Victory over Japan Day) and VE Day (or Victory in Europe Day). World War II had definable players and distinct winners and losers. Veterans came home to parades and accolades at the time and they, together with their fellow civilian Americans, went down in the annals of American history as the "greatest generation."[29]

American support for and understanding of the war in Afghanistan and Iraq differs greatly. The enemy was tangible after terrorists hijacked four planes on 9/11. The entire country had an emotional response similar to the reaction to the Pearl Harbor attack. Yet, the players were not as distinct as in the past, and the number of those who would be needed to fight were not as great either. Although the United States had more than 2.6 million men and women serve in Iraq and Afghanistan, this number is remarkably low when compared with the more than 9 million who served in the Vietnam War during a 10-year period (1965–75) when the population was between 194.3 and 216 million. The ratio for World War II is even bigger when considering that 16 million served during a four-year period (1941–45) at a time when the nation's population was between 133.4 and 139.9 million.[30] The percentage of those who served in the Afghanistan and Iraq wars was barely 1 percent of American society. So while nearly everyone knew someone fighting in Europe or the Pacific in the 1940s, very few had family or friends fighting in either Afghanistan or Iraq.

Though the public was supportive of the U.S. troops, they were not socially invested in the war. The cost of the war was added to the national debt.

---

[29] Tom Brokaw, *The Greatest Generation* (New York: Random House, 1998).

[30] Population statistics for the United States during these periods can be found at "U.S. Population by Year," http://www.multpl.com/united-states-population/table.

Without sacrifices being made on the home front, young veterans felt the American society slowly disconnected from the wars. Veterans saw a national apathy as those back home continued on with their daily lives, forgetting the wars and, therefore, the veterans who were risking and losing their lives daily in two vicious conflicts.

Americans may not wholly understand how powerful the experience of going to war is, or the feelings expressed by war veterans, but they should hear about the experience, while veterans need to share their stories for the sake of each other.

This book provides an understanding of what the experience of going to war was like for a selected group of veterans of Afghanistan and Iraq wars. The essays are written by several Marines and a U.S. Navy chaplain who deployed with Marines in combat. Though combat is probably the most powerful experience of being at war, many other aspects are also important when comprehending fully what going to war is like. Thus, several essays deal with the broader experience of being in a war zone and not necessarily with combat. Many books provide an excellent understanding of combat, this work provides insight on the impact war makes on individuals overall.

The essays in this book convey that going to war is a complex phenomenon that begins before troops even arrive in a war zone and lasts well beyond the end of deployment. Going to war makes an impact on individual servicemembers, their immediate family, their larger family, and their fellow Marines.

In putting this book together, we asked the authors to write their stories with a gritty directness, combining their personal and professional reflections to express the emotional impact of their wartime experiences. The authors' language and approaches can, at times, be blunt, and that is by design. As editors, we wanted to have the most accessible essays without filtering, interfering or, even worse, compromising the voice of those who have been at war.

---

[*] Paolo Tripodi is professor of ethics and ethics branch head at the Lejeune Leadership Institute, Marine Corps University.

CHAPTER ONE

CAPTAIN DOUG ZEMBIEC
BY SERGEANT MAJOR WILLIAM S. SKILES (RET)

Sergeant Major William S. Skiles served in the U.S. Marine Corps for 32 years. During his career, Skiles held a variety of positions including sniper instructor, reconnaissance platoon sergeant, Royal Marine exchange instructor, drill instructor, and Marine Corps rifle team shooter. He was decorated three times for his heroic combat actions during his deployment to Fallujah, Iraq, in 2004. The following story is focused on that deployment when he served as a first sergeant with Echo Company, 2d Battalion, 1st Marines. Skiles reflects on his relationship with company commander, Captain Douglas A. Zembiec, and on how Zembiec's leadership, character, and humility influenced Skiles's life, as well as the lives of all the Marines in Echo Company who saw combat together. Skiles received the Colonel Francis F. "Fox" Parry Writing Award for initiative in combat for his January 2008 Marine Corps Gazette *article, "Urban Combat Casualty Evacuation," which covered the same time period depicted in this essay.*

When the ramp drops tomorrow, who will come out? What type of human beings will come out? Will those humans honor their country, their corps, and more important, the Marines to their left and right? When the ramps dropped from the boats on the beaches of Iwo Jima, who came out? When the ramps dropped from the helicopters in Vietnam, who came out? Honorable warriors came out to uphold the virtues and values by their actions on the battlefield. This is our time and our legacy. This is our Vietnam. This is our Okinawa. Carry the honor and traditions that formed our great nation into battle. Climb aboard, Marines![1]

These were the words spoken by Captain Douglas A. Zembiec as he addressed his company of Marines before they entered the city of Fallujah, Iraq, in 2004. I was one of those Marines. Why were these words so important? Were these words a way to stop the hatred we felt against the individuals that killed four Americans the day prior? Was it because we had weapons in our hands and could take lives with one pull? Was it a reminder that regardless of what we all felt inside, our ethical and moral obligations to each other and this great country of ours should be paramount in our minds? By the captain's words and, more important, his actions, we entered and left a battle with a clear conscience and a feeling that we did good in the face of extreme adversity. Whenever he spoke to us, we listened. From my first encounter with him, he had a certain way to make me believe in his actions and words.

I arrived to 2d Battalion, 1st Marines, on 12 October 2003, as a senior first sergeant.[2] The sergeant major assigned me to Echo Company and paired me with Captain Douglas Zembiec. As I left the sergeant major's office, he said, "good luck with that warrior of officers!" I did not know how to take the comment, but soon found out what he meant.

---

[1] Capt Douglas Zembiec (speech, Operation Vigilant Resolve, Fallujah, Iraq, 3 April 2004), based on author's recollection.

[2] The 2d Battalion, 1st Marines, is based at Marine Corps Base Camp Pendleton, CA.

Though I did not know it at the time, meeting Captain Zembiec would change my life. As I walked into his office, he dutifully got up and offered the type of firm handshake I would expect of a Marine. He said, "First Sergeant Skiles, Bill, I've been waiting for your arrival, and we are going to do good things together." I reviewed his vision for the 139-man infantry company he commanded, and we discussed how I could assist him to prepare the company for battle. He knew at that point that we were headed to a place in Iraq known as Fallujah. It seems funny today to think that, at the time, we had no idea where Fallujah was; it was just a mysterious place on a map. Our unit was to depart in late February 2004 and arrive in Kuwait. We would then head north to relieve the U.S. Army unit that was currently stationed there. This was going to be the first time the Marines traveled back to Iraq since President George W. Bush declared the end of combat operations in May 2003. Captain Zembiec also briefed me on the training that he would like our men to do, which was not necessarily the type of training the higher-ups wanted us to provide. Somehow, he knew we were not heading into a mop-up situation and that we needed to be prepared for combat.

As I left his office, the captain gave me a reassuring handshake and invited me to the evening's social event at his house. He then asked me to bring my wife, Tanis, to this event so he could meet her. At that moment, I realized he had taken the time to find out my wife's name before meeting me; this is something that not all leaders in the Marine Corps do.

Tanis and I arrived at the captain's apartment in Mission Viejo, California. What was inside taught me the true meaning of strength through humility. After Tanis and I were introduced to others in the room, I wandered through the apartment, taking note of all the awards that were hanging on the walls. What Doug had accomplished was amazing. He graduated from the U.S. Naval Academy in 1995 and was a two-time NCAA all-American wrestler. He also had experience with force reconnaissance, serving as a platoon commander in Bosnia, where he saw combat face to face. Prior to coming to 2d Battalion, 1st Marines, he had graduated from resident

Expeditionary Warfare School (EWS) at Marine Corps Base (MCB) Quantico, Virginia, a school very few captains are selected to attend.[3]

Why was I so surprised to learn of all Doug's accomplishments? Mostly, I was impressed by his humility and the fact that he was never arrogant, despite his personal successes. If I had not visited his apartment that day, I may never have found out that he attended the Naval Academy or that he was a competitive athlete. It was as though he wanted to be just a Marine sharing an experience with other Marines without calling attention to his accomplishments. Humility was Doug Zembiec's greatest strength, and in the next few months of training for Fallujah, he demonstrated this attribute in every way possible.

From October 2003 to February 2004, our battalion was in predeployment status, and we trained. For the most part, the training we received was typical; we attended hundreds of briefs about what to do and what not to do during deployment, and we participated in various work-up exercises. However, it was Captain Zembiec's company training that made us lethal. Each morning, Doug would lead us through our physical fitness training regimen. Here are a few examples of Doug's humility in action during this training. Because of his physical prowess, one might think Doug would always strive to be the first to finish the company runs and that he would mock the Marines who were not able to keep up with him. This is the typical "lead from the front" mentality that our young leaders are encouraged to employ. I learned from Doug, however, that true leaders finish in the middle. "How can a leader see what's behind him if he finishes first?" he asked.

Unaware that Doug was an all-American wrestler in college, one of our bigger lance corporals challenged him to a grapple. Confident that Doug would win the match, I stood by to watch the spectacle. In a matter of three minutes, I learned again the lessons of humility from a captain of Marines, a true leader. He actually pretended to struggle and allowed the young

---

[3] EWS, one of several professional military education (PME) schools at Quantico, educates and trains company-grade Marine air-ground task force officers to serve in expeditionary environments.

MAJOR DOUGLAS ALEXANDER ZEMBIEC
14 APRIL 1973–11 MAY 2007*

Douglas A. Zembiec graduated from the U.S. Naval Academy in 1995 and was commissioned a second lieutenant in the Marine Corps on 31 May 1995. In April 1996, after finishing The Basic School and Infantry Officers Course, he was assigned to 1st Battalion, 6th Marine Regiment, as a rifle platoon commander. In June 1997, after successfully passing the force reconnaissance indoctrination, he was transferred to 2d Force Reconnaissance Company at MCB Camp Lejeune, North Carolina, where he served for two and a half years as a platoon commander, eight months as an interim company commander, and one month as an operations officer. As a platoon commander, his force reconnaissance platoon was among the first conventional forces to enter Kosovo during Operation Joint Guardian in June 1999.

In September 2000, Zembiec transferred to the Amphibious Reconnaissance School located in Fort Story, Virginia, where he served as the assistant officer in charge for two years. He was then selected to attend the Expeditionary Warfare School at MCB Quantico, Virginia, graduating in May 2003.

In July 2003, he took command of Echo Company, 2d Battalion, 1st Marines. As a rifle company commander, he led 168 Marines and sailors in the first conventional ground assault into Fallujah, Iraq, during Operation Vigilant Resolve.

In November 2004, Zembiec turned over his company command and served as an assistant operations officer at the 1st Special Operations Training Group (1st SOTG), where he ran the urban patrolling/military operations in urban terrain, and tank-infantry training packages

---

* Official Marine Corps biography of Douglas A. Zembiec as attached to Marine Corps Order 1650.55.

for the 13th Marine Expeditionary Unit in preparation for a deployment to Iraq.

On 10 June 2005, he transferred from 1st SOTG to the Regional Support Element for Headquarters Marine Corps. He was promoted to major on 1 July 2005.

On 11 May 2007, Zembiec was serving his fourth combat tour in Iraq when he was killed by small-arms fire while leading a raid in Baghdad. Zembiec was leading a unit of Iraqi forces he had helped train. Reports from fellow servicemembers present in the dark Baghdad alley where Zembiec was killed indicated that Zembiec had warned his troops to get down before doing so himself and was hit by enemy fire. The initial radio report indicated "five wounded and one martyred" with Zembiec killed and his men saved by his warning.

Zembiec's personal awards include the Silver Star, a Bronze Star with distinguished device, two Purple Hearts, two Navy and Marine Corps Commendations, two Navy and Marine Corps Achievement Medals, two Combat Action ribbons, a Navy Unit Commendation, two National Defense Service Medals, a Kosovo Campaign Medal, three Sea Service Deployment ribbons, a Global War on Terrorism Expeditionary Medal, a Global War on Terrorism Service Medal, a Humanitarian Service Medal, and a NATO Medal.

SILVER STAR CITATION
DOUGLAS A. ZEMBIEC

The President of the United States of America takes pride in presenting the Silver Star (Posthumously) to Major Douglas Alexander Zembiec, United States Marine Corps, for conspicuous gallantry and intrepidity in action against the enemy while serving as a Marine Advisor, Iraq Assistance Group, Multi-National Corps, Iraq, in support of Operation IRAQI FREEDOM on 11 May 2007. Attacking from concealed and fortified positions, an enemy force engaged Major Zembiec's assault team, firing crew-served automatic weapons and various small arms. He boldly moved forward and immediately directed the bulk of his assault team to take cover. Under withering enemy fire, Major Zembiec remained in an exposed, but tactically critical, position in order to provide leadership and direct effective suppressive fire on the enemy combatant positions with his assault team's machine gun. In doing so, he received the brunt of the enemy's fire, was struck and succumbed to his wounds. Emboldened by his actions his team and supporting assault force aggressively engaged the enemy combatants. Major Zembiec's quick thinking and timely action to re-orient his team's machine gun enabled the remaining members of his unit to rapidly and accurately engage the primary source of the enemy's fire saving the lives of his comrades. By his bold initiative, undaunted courage, and complete dedication to duty, Major Zembiec reflected great credit upon himself and upheld the highest traditions of the Marine Corps and the United States Naval Service.

Marine to pin him. As they both got to their feet, Doug embraced the Marine and said, "Great job, warrior. The enemy is in for it with strength like that." The young Marine smiled and walked away proud. Doug knew something the rest of us did not. He understood the importance of being comfortable in his own skin. He always told me there is no need for showboating when you are comfortable with your strengths and flaws.

Doug only lost his temper with me once, but even then, he acted as a professional. It was a typical early morning workday, and we all came together for a company formation. The platoon sergeants and company gunny gave me the thumbs-up to let me know that everyone was in formation, and I took their word for it instead of double-checking to make sure all of the Marines were present. I walked down to the captain's office to let him know that the company was standing by in formation. Yet my accounting of the Marines present was incorrect. Doug said a few words to the Marines of Echo Company and then met with the company officers, while I met with the enlisted. After the meetings, Doug called me into his office and asked me to close the door. When I turned around to talk to him, he was inches away from my face. He looked me straight in the eye and said, "Don't you ever lie to me again!" All I could reply was, "Yes, sir." He then turned and calmly asked, "What else do we have on the training schedule today?" Like a true professional, he did not allow his emotions to distract him from the mission. That day I learned two other attributes of leadership: accountability and passion.

One last highlight of our time together in predeployment training was during the regimental field mess night, at which all the Marines from 1st Marine Regiment broke bread, shared embarrassing stories from training, and drank alcohol until we all passed out in the brush. Over a few beers, Doug's compassion and concern for the lives of our Marines was obvious. He shed a tear here and there, talking about what we would face and of his hope of bringing everyone home alive. Waking up after this event is something I will never forget because the scene was similar to events we would witness as we battled in Fallujah later that year. There were bodies everywhere! The smell of alcohol was apparent, and trying to find my Marines was difficult. In fact, we just let them sleep until they got up on their own.

Before we left Camp Pendleton for Kuwait, Doug got the unit together and had a contest. He said that whoever came up with the best company call sign would get time off. In response, a young lance corporal came up with the call sign, "War Hammer." From then on, when we talked to one another on our radios, we would refer to each other as members of the War Hammer company. For instance, Doug answered to War Hammer 6 while I was referred to as War Hammer 8. On 29 February 2004, the Marines of War Hammer went to war.

Our flight to Kuwait was typically long with various stops and customarily poor service. We could not wait to hit the ground. We transferred all of our gear from the plane to buses and drove hours to our holding base to acclimatize. During the next couple weeks as we waited to make the long drive to Fallujah, Doug was the only company commander who had his Marines participate in daily physical training exercises. We also did company runs just to reinforce the idea that we were a tight-knit, well-oiled machine. I think other companies were jealous of the cohesion that existed between the Marines in our group. Doug and I always ate chow together, and we discussed, from time to time, our mission in this mystery city called Fallujah. Like all good Marines, we wanted a mission statement and commander's intent that would outline the purpose of our mission, but we were not receiving that type of direction. During this time, we attended meetings about unit discipline around the camp. We were provided with long lists of regulations regarding the type of behavior that was appropriate at the joint military camp. However, there was very little planning taking place. It was as though going to Iraq was not the priority despite the inevitability of deploying to that place.

As we prepared to leave the camp and head north into Iraq, Doug and I agreed that what would take place next was the hardest thing we had to do in our entire time serving together. Our company was instructed to drive 36 vehicles across hundreds of miles of unsecured territory in Iraq. We were expected to have numerous attachments, and all of the Marines and vehicles were to arrive intact within three days.

First of all, most government vehicles were not well maintained. We had 7-ton trucks, jalopy-looking humvees, dump trucks that looked like they were from World War II, and fuel trucks that were already riding on spare tires.[4] Doug and I did our best to organize this disaster and to develop security plans, breakdown plans, and troop issue plans. It was Doug's passion and courage that ensured the success of this convoy from hell.

As we were driving to Fallujah, a U.S. Army unit with five vehicles decided to integrate into our convoy. Doug told the driver of our vehicle to speed up close to the Army unit and find the leader of this intrusive force.

---

[4] Humvee, or High Mobility Multipurpose Wheeled Vehicle (HMMWV). Also known as a military hummer.

Our driver did exactly that. He cut off the Army vehicle, running into its side. A soldier got out of the Army vehicle and approached Doug, and immediately, a heated argument ensued. After a lot of shouting, the soldiers got back in their vehicles and stated they would continue to follow us, regardless of the concerns Doug had expressed. I could not believe what I saw next. As the soldiers attempted to drive away, Doug stood in front of their vehicle. The soldier then began to drive over Doug, so he leaped on the hood of the hummer and held on to the windshield wipers. The hummer took off, and we followed it for about a quarter mile before the soldiers pulled over. To my amazement, Doug was still attached to the hood! Even before the valor of combat, Doug showed his men that he was willing to stand his ground, no matter what the personal consequences.

Camp Baharia, located three miles from downtown Fallujah, was to be our home for the next seven months.[5] When we arrived, the Army unit we were replacing taught us about the local populace, the terrain, improvised explosive device (IED) lanes, and areas to avoid. Doug and I both thanked the leaders of the unit for a great turnover and their professionalism, a welcomed contrast to the treatment we had received from the soldier who had attempted to run Doug over with the hummer.

It was not long before we had our first hostile encounter. On one of the last days of turnover from the Army, we were in downtown Fallujah conducting some grip and grins with the local town leaders.[6] I was on top of a tall building, along with about 100 other soldiers and Marines. We were keeping watch and providing security in case there was an incident during the meetings. Doug was situated near the vehicles and was chatting with another member of our company. Without warning, a mortar round hit dead center in the middle of the company's compound. Then, two more mortars hit around the compound, and one exploded where 10 Marines and soldiers were gathered. As chaos ensued, I had the chance to witness Doug Zembiec in action when lives were on the line.

---

[5] Camp Baharia, also known as Dreamland, was located on the outskirts of Fallujah, Iraq.
[6] Grip and grin (handshake and smile) refers to the general meet and greet that would take place between military units and town locals.

Through my binoculars, I watched him leap over a high wall and run out to where 10 troops lay on the ground. All of them were crawling and were clearly injured by the mortar explosion. Without hesitation, Doug climbed up the stairs and carried the more severely wounded over the wall to safety. He made numerous trips, and only one other Marine helped get all 10 servicemembers back over the wall. Once the wounded were all together, some of our vehicles left to find medical help. By now, all of us on the rooftops were being shot at by small-arms fire, so we hunkered down and slowly made our way to the vehicles. Once all were accounted for, we drove out as well. I assisted at a point outside the city to render medical assistance as best I could. By the end of the day, 10 men were hit. There were no deaths, but the Marines accounted for 5 out of the 10 casualties, and it was only our tenth day in the country. As things calmed down, I remember Doug and I walking to the chow hall with blood on our boots and trousers. All Doug could tell me was that "it's going to be a long seven months, and we better get a plan for more of the same." When I got back to the hooch that I shared with Doug, I started writing down the heroics of this captain of Marines, mostly to never forget but also to tell everyone about this man and to be an eyewitness to greatness in the face of adversity.[7]

During the next seven months, Doug's words proved to be true. We saw too many skirmishes and heated exchanges of battle with the enemy to recall all of them, but I will try to capture the character and heroics of Doug Zembiec that stood out to me and all the Marines of Echo Company.

On 26 March 2004, the warriors of War Hammer headed into the city of Fallujah on foot. This would be the unit's first true test of discipline in battle that Doug had prepared us for. The day before, a lone Marine was killed in the city, and our battalion's response to this aggression was to send in a couple Marine companies to go fishing for insurgents. That day, alongside Doug, we battled insurgents street by street, block by block, and we controlled about one-third of the city—sharing this moment with Doug was an experience beyond description. His demeanor and patience under fire made the rest of us more confident. He never fired a shot but made his

---

[7] *Hooch* is the term for a military dwelling during deployments.

way down the streets, coolly and calmly using his radio to keep control of the three platoons. He was moving so quickly that the lance corporal radio operator was having difficulty keeping up with the good captain. It was only when we got word that one of our own had been wounded that a look of panic hit his face. I told him, "I've got this," and "You keep the others in the fight." He just gave me a thumbs-up, and I ran toward the area where the injured Marine was.

This Marine was shot in the hip and needed to be taken to Bravo Surgical Unit aboard Camp Fallujah, our main headquarters.[8] I flagged down one of the Weapons Company vehicles and asked if they could assist. I took a door off the hinges of a hut nearby and placed the injured Marine on it. We then stuck the door and Marine in the back of the vehicle and headed back to our camp. This was the first of the 78 casualty missions that I went on for Doug and our Marines. After dropping off the Marine at medical, I went back out to rejoin the fight. I found Doug and gave him the thumbs-up to let him know the Marine was taken care of.

After this daylong battle, Echo Company fell back. We dug in around an area known as the cloverleaf.[9] The next day, we learned that we had killed more than 20 enemy combatants, yet we had one Marine wounded. We waited for a counterattack that never came. After two days, Doug was frustrated by the lack of information we had gained, but he still felt that it would be foolish to just give up the ground. On the third morning, we were ordered back to our home, Camp Baharia. This is when Doug and I had a serious talk about what each of us would be responsible for. "Having one of our Marines down with no real medivac plan was stupid and reckless on our part," Doug insisted. Then and there, we decided who would take the fight to the enemy and who would be responsible for the fallen.

---

[8] Camp Fallujah, MEK Compound was a military base located inside the city of Fallujah.

[9] The cloverleaf interchange was a military entry control point for Fallujah located just east of the city at the intersection of Iraq's Highway 1 (Main Supply Route [MSR] Mobile) and Iraq Highway 10 (MSR Michigan). See Timothy S. McWilliams, *U.S. Marines in Battle: Fallujah, November–December 2004*, with Nicholas J. Schlosser (Quantico, VA: Marine Corps History Division, 2014).

Doug made sure I knew his appreciation for my efforts, and I would like to share what Doug wrote on my going-away plaque that he gave me on 12 December 2004, when we came home to the states and I was promoted to sergeant major. It reads,

> From the Marines of War Hammer, you slayed the enemy when needed, and took care of our wounded, and for that we are brothers forever. May the memories of this company stay with you forever and ever and it's been an honor to share the experience of valor and sacrifice with you. War Hammer 8, I salute you.

Things that Doug gave me and wrote for me, and items that we shared, are still in my heart and soul today.

As the month of April 2004 came, so did the sacrifice. After the American contractors were slain in the streets and hung on the bridge, we went back to this city of hell to stay for the entire month.[10] What occurred during this month affected us, but most of all, it reinforced what Doug Zembiec instilled in us and how we knew things would work out so long as he was in the fight.

Echo Company fought the insurgents daily, and thank the Lord, they were lousy shots. We occupied houses on the edge of the city in an area known as Jolan Heights.[11] We built up our positions, ready for a counterattack at every waking moment. The beautiful thing about having Doug as my company commander was that he always had time for me. Between firefights, we both agreed to sit down and document the heroics of our Marines that day. We focused on recording eyewitness statements that would hopefully reward these young Marines for their fearlessness in the face of the enemy and give credit where credit was due. By the end of our seven months in Iraq, our company had written more valor awards than all

---

[10] On 31 March 2004 in Fallujah, insurgents killed four employees of the security contracting company Blackwater, and two were hung from a bridge in the city. See McWilliams, *U.S. Marines in Battle.*

[11] Jolan Heights is a residential area in northwestern Fallujah. The district is the oldest area of the city. See McWilliams, *U.S. Marines in Battle.*

other companies combined.[12] It was not because we saw more combat; it was because our company commander had the humility to remember daily the Marines in front of him and to make sure their stories were told. Doug got upset if a platoon commander put off writing about a Marine due to the commander's fatigue, because he truly enjoyed writing about the heroics of his men and wished to write a book about them someday. In the same light of valor, we also had those who made the ultimate sacrifice. Doug confided in me a lot about his experience writing condolence letters to the families of our dead Marines. Knowing a grieving mother and father would receive these letters, he wanted to make sure they were the best he ever wrote. All in all, Echo Company had more than 78 dead and wounded Marines out of a company of 139. Some of these warriors were hit by enemy fire twice and still continued to fight. This is a true testament to how Doug Zembiec inspired his Marines through his leadership and humility.

In honor of his heroics in battle, I wrote-up seven separate, heroic-action awards for Captain Doug Zembiec. I had hoped one of these write-ups would make a difference. After all, if I did not record the heroics of my captain, who would? In fact, one event I described in one of these write-ups is portrayed on a statue that Marine Corps Special Operations Command (MARSOC) gives annually to its team leader of the year. Named after the brave captain who taught me so much during our deployment, this annual award became known as the Zembiec Award.[13] Below is the citation I wrote to recognize Doug for his heroic actions:

> April 6, 2004, Captain Doug Zembiec went above and beyond the call of duty. While in a static, defensive position, Echo Company came under a mortar and small-arms attack by a platoon sized insurgent element. Two U.S. M-1A [Abrams] Tanks were in support and as Echo received the fire, the two tanks went forward into the outskirts of the city laying supporting fires. The two tanks were using their main gun and machine

---

[12] Awards recognizing Marines for valorous actions include the Medal of Honor, the Navy Cross, the Silver Star, and the Bronze Star with "V" for valor.

[13] The Douglas A. Zembiec Award for Outstanding Leadership in Special Operations was first awarded in 2011.

gun to repel hostiles. Doug Zembiec noticed that the two buttoned up talks were firing at the wrong building, so he leaped forward, ran over two walls, and exposed himself to enemy fire as he approached the rear of one of the tanks. The tank phone was not quite working, so Doug had no other choice but to get on top of this moving and firing tank to try to get the commander's attention and redirect their fire to where the insurgents really were. As enemy bullets ripped all around him, Doug banged on the hatch of the tank and the commander opened the hatch as to communicate with Doug. Doug used his knife hand to show the commander which building the enemy was in. Doug then took a magazine of tracers and fired these visible bullets into the building where the tank should be firing. As I looked on, the tank slowly rotated its main gun, and with Doug still on the back, the tank fired its main gun into the correct house. The intense blast and shock knocked Doug off his feet. Doug gave the thumbs-up to the commander and ran back to our lines. Due to his heroics, the bad guys quit firing and retreated as the tank finished off a couple more. The witnessing Marines all gave our Captain a high-five and there were smiles all around.[14]

In October 2004, the warriors of Echo Company returned home. One of the proudest moments that Doug and I shared was when we got off the buses, formed the men up, and marched toward our waiting families. I remember seeing Pam, Doug's fiancée, jumping into his arms and almost taking him down. We looked at each other, and I said, "See you tomorrow." Then, in December, Doug and I parted ways. I was promoted and reassigned, and Doug had orders as well. We stayed in contact throughout the following months, and in mid-2005, I was honored to attend Doug and Pam's wedding on the U.S. Naval Academy grounds. I toasted them and told some of the stories that Doug and I shared in our hooch in 2004. I spoke of how hairy this man was from the eyebrows down and of his ob-

_____

[14] Author's notes on Douglas Zembiec's actions of 6 April 2004 (submitted to the Marine Corps for valor award). The Marine Corps awarded Zembiec the Bronze Star with "V" on 8 December 2004 for his actions in Fallujah between 6 April 2004 and 1 May 2004.

sessive need to keep his toenails clipped. I had a great time remembering the fun we had during the most difficult, but most rewarding, time in our lives. After the wedding, we lost touch for a bit. I learned that Doug volunteered for duty with special operations for the Marine Corps. The missions he was selected for are still unknown; what little I do know about what Doug was doing in Iraq I learned from Pam. The twelfth of May 2007 is the day I wept.

I was a sergeant major of a squadron of Marines, and I was doing my daily routine when I received a phone call. One of my former Marines from Echo Company asked if I had heard the news. I asked, "What news?" He replied, "Doug's dead." I wanted what I had just heard to be a bad joke. I sat down and asked, "How did he die? How do you know?" He said Doug had died the day before, but the Marine Corps had just received the news. All I could do was try to breathe. There was no way a goliath of a man like Doug could be dead. I said my thanks, hung up the phone, and sat silently for a few minutes. And then it came. I have never in my life experienced what I went through that day in my office. Rage and tears flowed to the point that I could hardly catch my breath. I wept for at least an hour before I composed myself enough to call my wife. I went home early that day and stared at nothing. Then, I received a call asking if I could fly to the Naval Academy to assist with Doug's funeral.

I left that night for Annapolis, where I met up with others who were also shocked to hear the news of Doug's death. Despite our grief, we knew we had to regroup to do this deed together. I shared the news of Doug's death with the folks I met that day, and I think we all felt sorry for those who were not fortunate enough to have met Doug. We considered ourselves better for having been in his presence at one time or another. The things Doug's friends and family said about him during his funeral will always stick with me. In a story recounted at Zembiec's funeral and reported in coverage by the *Washington Post*, Zembiec and his father, Don, were driving on to Camp Pendleton and were stopped at the gate by a young Marine. "Are you Captain Zembiec's father?" the Marine had asked. "Yes," his father said. "I was with your son in Fallujah," the Marine said. "He was my company commander. If we had to go back in there, I would follow him with a spoon."[15] Almost the entire Echo Company gathered at the Arlington National Cemetery, Virginia, that day to say goodbye to the "Lion of Fallujah," as we called him.

Doug told me one time that he really looked up to me and appreciated how I handled the Marines and their personal problems. It is funny to reflect on this conversation that I had with Doug, because for the rest of my life, I will look up to him. The following is what I wrote the night I heard about Doug's death. It was originally sent as an e-mail, but it went viral to blog sites. It is the best way to conclude this essay about the life and work of Doug Zembiec:

I'm SgtMaj Bill Skiles, and I was Doug's 1stSgt in Echo Co. in 2004 in Fallujah. I would like to tell you about the Doug Zembiec that you won't read about in papers. I shared a hooch with this man for seven months and we would talk about everything from his Marines to what it will be like to be married. Doug was known for his tremendous warrior spirit and his physical strength. He was a physical specimen, but he had a heart of gold. The qualities that I still live with thanks to him are humility and sincerity, Doug would be the first to hug a PFC and tell him it's OK, instead of putting him down for being weak. He would be the first person to stand up for you if he felt you were being treated unfairly. When he told someone he would do something, he did it and made sure you knew the results. He wouldn't sleep until he knew you understood what was happening. Doug was confident is [in] his own abilities, and he had every right to be the most arrogant man alive. But he knew who he was and would always tell me that any leader who had to be arrogant towards his own Marines was probably thin skinned and insecure. He would call some of these Marines "junior varsity."

Doug and I made a deal on the day our first wounded went down in late March '04. The deal was that I would take care of and account for all wounded and he would keep the rest of the Marines focused on

---

[15] Dan Morse, "Salute to a Memorable Marine," *Washington Post*, 17 May 2007, http://www.washingtonpost.com/wp-dyn/content/article/2007/05/16/AR2007051602860.html.

the fight. This agreement was made because he could not handle seeing his Marines bleeding and hurting. . . . He and I would weep behind closed doors during some of the trying times with mass casualties. Doug's emotions were always worn on his sleeve, and I really admired that. His troops admired that. . . . He showed us all that he was human; he cared deeply about us and felt what we felt. ALL of his troops would have given their lives for Doug if needed; I cannot name another commander who could say that. He wasn't fake; he wasn't the most politically correct officer, but the troops that walked the streets with him and fought and sacrificed with him understood. That bond is hard to teach any ego. . . . I wish all commanders could learn just a little of the humility and sincerity this warrior displayed daily to every Marine, regardless of rank. Doug's Marines loved to laugh with him, cry with him, and mostly, to fight and kill the enemy with him. Every Marine knew that when Doug showed up to a fight, things were going to be ok.

Doug allowed the chaplain to perform services during firefights, comforting our grieving warriors after loss. He listened to our corpsman about how to take better care of the fallen. From his firm handshake to a grieving hug, I will miss him until I join him. I will miss this man, the hairiest man on earth from the eyebrows on down. The poor guy had no hair above his eyebrows, but he was a human woolly pulley every where [*sic*] else. He would try to shave his back before patrols and always missed various spots. And yes, I would help finish the job. What are buddies for? Doug Zembiec would never talk about himself, about what's he done, about any of his accomplishments, because he told me that no one really cares about what you have done. As you command, the Marines want to know what you can do now and [in] the future. Well said! The day Doug received his Bronze Star with "V", he wept. I wept and I hugged this warrior and no words were spoken . . . I know why we wept. We would talk over and over again about how with valor comes sacrifice, but he thought this valor medal would never match the sacrifice that his Marines went though [*sic*]. Humility again shows itself.

About his new family: Doug LOVED Pam and being a dad made him even more humble. His daughter's birth was the proudest day ever for him. Until that day, he told me the proudest moment in his life was leading the Marines of Echo Co in battle. I could talk for days about how much this man meant to me and to his Marines, but I know this man was the definition of what a Marine should be, what a committed husband and father should be, and what this country looks for in a true hero in every sense of the word. I will spend a couple hours with him tomorrow night when it's my turn to watch over his body, and we will finish what we talked about for those seven months.

I love you, Doug

Sgt Maj Bill Skiles[16]

REFLECTION POINTS

- How does a Marine, or a soldier, assimilate the need to be lethal—which requires a sense of brutality—while also having an ethical base? What role, if any, does the leader play in this process?

- How do the actions of leaders affect those who follow or lead alongside?

- Can leaders be chosen or trained to achieve the success Zembiec did with his men and women?

---

[16] Author's e-mail, 12 May 2007.

CHAPTER TWO

THE UNSPOKEN LEADERSHIP CHALLENGES OF COMMAND
COLONEL BRIAN S. CHRISTMAS

Colonel Brian S. Christmas, a Marine infantry officer, has been led and mentored by some of the best officers and enlisted military leaders throughout his childhood and career. Exposure to these professionals enabled him to accomplish missions alongside Marines, sailors, soldiers, airmen, but also interagency and international partners during multiple wartime deployments, including with the 24th Marine Expeditionary Unit (24th MEU) as a platoon commander with Battalion Landing Team 1/6 (BLT 1/6); during a unit deployment program (UDP) to Okinawa as a weapons company commander; as a 22d MEU BLT operations officer conducting combat operations in Uruzgan Province, Afghanistan; as a battalion executive officer conducting combat operations in Fallujah, Iraq; and as a battalion commander of 3d Battalion, 6th Marines, conducting combat operations in Marjah, Afghanistan. Colonel Christmas provides insight on leadership challenges and the importance of personal and unit preparation associated with combat losses on the battlefield. He also conveys to the reader that he holds his Afghanistan and Iraq campaign medals in the highest regard.

In 1991, during my junior year of college, I walked into the Marine Corps recruiter's office and declared my desire to serve my country as a U.S. Marine. To those who know me superficially, this career would have seemed like the natural choice. Both my father and my grandfather are well-known, highly respected Marines. My brother was a Marine platoon commander who actively served in Operations Desert Shield/Storm. My longtime girlfriend's father was also an esteemed active-duty Marine aviator who served in Desert Storm; his father-in-law served on Iwo Jima, Japan, during World War II, coincidentally on the same beach where my own grandfather served gallantly. I had a great-aunt, uncles, a godfather, and close friends who were all Marines. Given my lineage and the significant influence the Marine Corps had on my life, choosing it as a career seemed logical, even predictable, to the casual observer. But for those who knew me well, my decision was a complete surprise. To them, I was academically minded and prepping myself for law school and perhaps involvement in Washington, DC, politics. I had never expressed any interest in becoming a Marine or joining any other Service. They wondered what prompted this change of heart.

My decision was not an epiphany that happened overnight. It was based on the realization that I was not ready for law school. I lacked the desire and discipline, and I knew that if I was going to succeed, I would need to mature, to grow up if you will. I enjoyed college, not only for my studies but also for my role as the basketball manager. My basketball coach demanded excellence and commitment, not just from his players but also from his manager. Every person under the coach's supervision had a vital role to play in making the team a success. Our commitment earned us responsibility and trust. I was given authority and represented the organization. I made crucial decisions, interacted with others, and executed tasks based on a known mission and my coach's intent. The more I accomplished, the greater his confidence in my abilities grew, which was directly reflected by the complexity of each new task that I was assigned. I thrived in this environment and desired more.

My affinity for serving, leading, and making a difference, combined with my goal to mature and reset myself, led me to the Marine Corps' officer selection officer (OSO). I could not think of a finer institution that specialized in leadership and discipline. The numerous examples of Marines that I held in high regard were evidence enough. So, I developed my plan. I would attend Platoon Leaders Class (PLC) during the summer while continuing to maintain a heavy course load to graduate a semester early my senior year. After that, I would graduate from The Basic School (TBS) and serve my country.[1] After spending some time in the fleet, I would reevaluate my circumstances and consider applying for the Funded Law Education Program.[2] If all went well, I would attend law school and serve my additional commitment while gaining valuable legal experience in the military judicial system. I would then leave the Marine Corps with a heightened sense of discipline and purpose and return better equipped to pursue a legal profession and possibly "Beltway," or DC, politics. In my various visions of the future, never did I imagine the path I would actually end up taking.

It all started at TBS where officers are assigned their military occupational specialty (MOS). I chose infantry not because of my lineage but because I realized during the training and exercises that I genuinely enjoyed what I was doing and the leadership responsibilities associated with being a grunt. I was given orders to 3d Battalion, 8th Marines, at MCB Camp Lejeune, North Carolina. When I contacted my new unit, Lieutenant Colonel Christopher J. "Chris" Gunther, the battalion commander, informed me that I had two choices: I could go on leave (I had just gotten married and was prepping for the move to North Carolina) while the rest of the unit conducted mountain warfare training in California, or I could cut my leave short and deploy for a month to California with the unit. I knew the best way to get to know the unit and to gain the unit members' confidence was to prove myself to them in

---

[1] TBS is a Marine Corps training command, which trains and educates newly commissioned or appointed officers.

[2] The Funded Law Education Program provides an opportunity for officers to laterally transfer to the occupational specialty of judge advocate, performing legal services for Marine or Navy commands or organizations.

the field. So, after consultation with my wife, whom I had just returned with from our honeymoon, we decided to head to North Carolina immediately. A couple of weeks later, I was on an airplane with the rest of my battalion.

The trip to Bridgeport, California, taught me a great deal. I remember approaching my company commander on the way up Hill 9494 in the Mountain Warfare Training Center area in Bridgeport.[3] I had been a rifle platoon commander for less than a month. I expressed my concern for the Marines' ability to make the hike at the current pace with the extra weight of the winter combat load. Captain Robert F. "Bob" Killackey, the company commander, assured me, "They don't have a choice. They have to." I never imagined the admiration and sense of brotherhood that I would feel for my Marines and sailors who did exactly what was expected of them—they made it. They always made it. They do what they have to do. They rise to the occasion.

More than two decades later, Marines still amaze me, and my love for them remains steadfast. It is hard to imagine that I ever considered leaving them. I became a Marine because I realized that I needed what only the Marine Corps could offer. My desire to make a difference in the world while maturing and satisfying my penchant for leadership and responsibility was more than met by the Marine Corps. It raised the bar entirely. The Corps has groomed me during the past 20-plus years to realize that leadership and responsibility require trust, confidence, humility, and resilience. I learned these lessons at the platoon, company, battalion, and staff levels. Consequently, I acquired the character traits necessary to be successful. Leading also means serving the Marines and sailors in a manner that is effective and empowers them to rise to the occasion and make it happen like they always do. The joys of leading these young men and women in their quests to achieve their personal best is rewarding beyond words, but it is not without its own set of difficulties and heartbreak.

As a Marine, the greatest challenges came when I was in command. Such challenges, while occasionally discussed in the classroom as hypotheticals, are never truly understood until a person actually assumes the role of com-

---

[3] Hill 9494 refers to the highest elevation Marine units climb while conducting cold weather mountain warfare training in Bridgeport.

mander. Walking in those boots is an honor and privilege like no other. A hallowed relationship is forged—a deeply rooted connection and trust with the Marines, sailors, and families of the battalion you serve. This reality became clear to me when I took command of 3d Battalion, 6th Marines, in January 2009. After the unit's change of command, I told my wife that I could actually feel the weight of the colors as they were handed to me and I returned them to the sergeant major during the ceremony, which symbolized the battalion transferring from the hands of the previous commander to my own. I was not referring to their physical weight but rather the weight of responsibility a command brings. I did not know it at the time, but I would come to understand these unique challenges almost immediately.

During my first night in command, a Marine was killed in a motorcycle accident (at a time when motorcycle safety was paramount in the Marine Corps).[4] I had dealt with both garrison and combat losses in the past, but there is a significant difference when one experiences it as a battalion commander. The loss of any Marine or sailor is devastating for all members of a unit, but as the commanding officer, there is a deep feeling of responsibility that is almost parental in nature. As the one in charge, you mourn the loss alongside your battalion, but you must also move the unit forward; help the Marines understand what happened and what can be done to prevent future loss of life. This tragedy was the first during my time in command; regretfully, it would not be the last.

The lessons I learned during my command were many and did not fall into one neat category; the lessons were diverse and ranged from unit readiness and family support to battlefield problem solving and casualty assistance. Some lessons I anticipated and prepared for; others, I became conversant with on the job, and still others I lived in the moment. The most important lesson was also the most difficult, because it involved the harsh reality in which no one wants to be proficient. The loss of a comrade in arms. The proper notification and subsequent support of family members. The mourning of the battalion family. The need to grieve within the unit and then

---

[4] The Naval Safety Center reported 25 fatalities for the Marine Corps in 2008, meaning more Marines died that year in motorcycle accidents than were killed during combat in Iraq.

move forward with the mission. A battalion commander's biggest challenge is his fallen Marines and sailors. Some of the necessary points to consider are detailed in the paragraphs that follow in hopes that readers will consider, mentally prepare, and explore their own solutions to similar scenarios. It is essential for a battalion commander to be forward thinking, anticipating and preparing a unit to handle anything, even death. The Marines and sailors deserve it. Their families deserve it. The fallen angels deserve it.

Uncertainty is the enemy of preparation. It attempts to undermine even the best-laid plans. A few months after I took command, the battalion was conducting extensive training at the U.S. Army's Fort A. P. Hill, Virginia, in preparation for the unit's deployment to Iraq as part of Operation Iraqi Freedom. I was informed that our departure date was now ambiguous and that the battalion might not deploy at all. Proper training requires intense focus and motivation by those preparing to deploy for combat. Now, in the middle of the unit's training, I had to distract the Marines and sailors with the news that our mission was in limbo. This announcement had a terribly negative effect on morale.

Then a few weeks later, the decision to deploy was made but with a location change to Afghanistan instead of Iraq. The battalion was given an abbreviated 30-day timeline in which to pack out and depart. Rapid response and compressed timelines are scenarios Marines always prepare for and handle well, but those scenarios are not something family members welcome or appreciate. Family separation is one of the hardest parts of this profession, and unexpected departure plans only increase the emotional intensity. It is a reality that must be addressed and taken as seriously as the pack out itself. Servicemembers need to know their families will be all right and well taken care of in their absences. Leaving without this assurance will cause the servicemembers to be distracted, which poses a real danger to themselves and to the battalion.

Throughout my career, I was exposed to different challenges, as well as the various kinds of leadership required to overcome or mitigate them. As a result, I had made a point of preparing myself for the opportunity to lead young Marines and sailors into combat; and when the time came,

I quickly realized there was even more involved in unit readiness than I initially thought. I needed to ensure that the details were taken care of not just for myself and the men and women serving alongside me, but for their family members too. I had no doubts that our Marines and sailors would be successful, because we were well prepared for combat. I ensured this to a fault. I also found that mastering the maneuvers of combat is critical, but other aspects of war are more difficult to prepare for in advance.

Preparation even when difficult is still key. PITMOV is an acronym that I learned from Hal Nunnally (1939–2004), the infamous basketball coach from Randolph-Macon College in Virginia. He explained to me that in everything we do—studies, sports, work, life—preparation is the mother of victory (PITMOV). When I joined the Marine Corps, this lesson was reinforced by the six Ps (proper preparation prevents p——s poor performance) embedded in our leadership traits and principles. It was a mantra underscored as I worked for and with such leaders as Lieutenant Colonel Bob Killackey, First Sergeant Al Ashbolt, Colonel Chris Gunther, Sergeant Major Mark Freres, now-Major General Paul E. Lefebvre, Master Gunnery Sergeant Alvin McNeal, Colonel David C. Fuquea, Lieutenant Colonel Asad Khan, Colonel William M. Jurney, and many others who all made certain that their subordinate leaders and young officers mentored by staff noncommissioned officer (SNCO) leadership understood and acted on this fundamental principle. It is no surprise then that, as a battalion commander, I put my heart and soul into ensuring our Marines and sailors were prepared for the worst challenges of combat. At times, it appeared as if this would be detrimental to my career.

I clearly articulated my vision, but garnering enthusiastic support of that vision from the families and my superiors was not always easy or successful on the first attempt. Once a plan was in place and the Marines and sailors understood the mission, they were all in. Their families, on the other hand, often did not recognize the true value of the time invested in training until the unit's return from deployment. As a commander, I understood that I needed to provide realistic and challenging training that my Marines and sailors understood and desired. When the goal of the training was clear, organized, and effectively executed, the Marines and sailors fully appreciated its value. I minimized transitions between tasks, so they were not standing around with nothing to do. I made use of every moment possible, but did not forget to let them breathe.

The schedules that I created allowed the servicemembers some time for themselves and for their families. It was never perfect, but it was done in a variety of ways to achieve an acceptable balance. The difficulty came in creating the same balance for the families. I tried to assist them in understanding the plan, recognizing their roles and their importance to our unit readiness and overall success. I have always made the analogy that our families are like those that stand behind the spartan line.[5] From the home front, family members are the ones who have their hands figuratively placed in the small of our backs, pushing firmly, providing us the much-needed support to ensure that we do our job. They may struggle to provide this crucial support, especially if they do not feel like they are a part of the plan.

For this purpose, I attempted to include families from the beginning. They were now a part of my family and were treated as such. It was not always easy, but it was essential. My phone rang and my e-mail buzzed with numerous inquiries from my new family members. On occasion, they voiced opinions or concerns that contradicted my plan (something I desperately wanted it to become—our plan) or my feelings on the matter. I heard their concerns and delegated the handling of those concerns to the appropriate individuals in the chain of command.

To bolster the families' strengths and build their resiliency, they were introduced to outstanding programs established through Marine Corps Community Services (MCCS), the Family Readiness Program, Military One Source, and other military programs that provide support services during a family member's absence. If the issue required my personal involvement, I answered it myself. I then made note to follow up on the initial action. It

---

[5] Scholars argue that the formation of the ancient Greek phalanx incorporated Spartan soldiers who physically pushed the military's front line forward. See Victor Davis Hanson, *The Western Way of War: Infantry Battle in Classical Greece* (New York: Knopf, 1989).

was crucial to complete these tasks and respond. I did not always provide the answer that the family members wanted to hear, but they always received a response. They knew that their voices were being heard and that the commanding officer was involved. This proactive approach fostered trust and confidence in me, my staff, and my subordinate commanders. Securing this trust and confidence prior to the unit's combat deployment was critical.

Despite my best efforts to smooth the way for the families that would be left behind, I sometimes questioned if I had gained enough of the families' trust and confidence prior to the deployment. I had an excellent Family Readiness Program.[6] I did my best to keep family members in the loop and involved, although I could not always provide them with what they wanted—time with their loved ones—because I took advantage of every training opportunity I could to get the battalion ready based on my assessment of what the unit needed to do. The better prepared the Marines were, the fewer our losses would be. I provided the Marines, sailors, and families with realistic, challenging, and time-consuming training prior to the deployment. Some people just could not see the need to train to the degree that we did. So, I was relieved and very appreciative when numerous spouses and parents approached me at the unit's deployment homecoming and thanked me for preparing both their Marines and their families. Then, they understood. I only wish that I had been able to convince them of the value of the training from the very beginning. War is hell, but preparing for hell is harder, especially when it comes to convincing everyone involved how necessary and important the preparation is.

Having said that, preparation can sometimes come with its own repercussions, as I soon discovered. The S-1 officer, the family readiness officer (FRO), and Key Volunteers were to deal with casualty notification.[7] It is heartbreaking for everyone involved, but it is also imperative that casualty

---

[6] Christmas's unit also instituted the strengths of the former Key Volunteer Network, which was led by spouses and active-duty leaders within the organization to provide an information support network for families.

[7] S-1 refers to the office managing personnel for the unit.

notifications be handled with the utmost care and competence. Indeed competence comes with preparation, but that is particularly challenging when it comes to the death of a loved one. I would learn this firsthand when my efforts to test this crucial notification system caused emotional turmoil for my own family.

While training at A. P. Hill, I spoke to my wife, Nikki, and asked her if she would agree to receive a mock phone call notifying her that I had been shot. She did not like the thought of it but understood the value of verifying that the notification process worked efficiently. She thoroughly prepared to grill the S-1 officer and the FRO with questions and concerns and to be as emotional as possible. With only my executive officer in the loop, I was "shot" by a sniper following a meeting with foreign role players. My jump platoon reacted with amazing vigor.[8] The Marines quickly recovered me and secured the area. Somehow, I seemed to be dragged through every puddle, every hole, and every bump. "Doc," the platoon corpsman, immediately dressed my wounds and directed the Marines for my movement to the vehicle. With the medevac request called in and the movement route defined, Sergeant Dennis K. Derr II moved us down the road while the battalion aid station (BAS) prepared for my arrival and the S-1 officer wrote up the casualty report.

Our excessive speed through A. P. Hill required me to come out of character, "Sergeant, great job, but slow down. The Army military police are not read into this role-play scenario." It was a smooth transition from Doc's care to that of the battalion surgeon and his team in the BAS; information flowed from the BAS to the S-1, and once ready, the casualty report was sent.

Upon notification, my wife began to ask the FRO a litany of questions, and then, the unexpected happened. My 5-year-old son walked out of the adjoining craft room, heard "shot" and "dad" and urgently returned to the craft room to share the upsetting news with his intense, Irish-blooded 8-year-old sister. My alarmed daughter authoritatively stomped over to her mother, grabbed the notebook out of my wife's hand, read the contents,

---

[8] A jump platoon ensures the safety of the battalion commander during missions.

and then screamed, "Daddy's been shot?" My son's reaction was natural for a boy. He proceeded to repeatedly strike my wife's legs in fear and anger as tears rushed down his face. Nikki hung up the phone and calmed our children. She later informed me, in Edgar Allen Poe style, "Nevermore, Brian! Nevermore!" The unintended impact on my family was not desirable, but it was a great test for the battalion as the unit would conduct these procedures more than 150 times during our time in Marjah, Afghanistan, including one notification to my own family due to an IED blast.

Preparing Marines and sailors for combat is something the Corps typically does well. When it comes to readying families, the task is much more complex and delicate, but no less necessary. My greatest obstacle was preparing the families for the possibility that any of the unit's Marines, sailors, Coalition partners, and interpreters could die. The only real method of doing so was to discuss such a risk, to ensure that everyone understood the spectrum of likely reactions, and to develop and emplace, as much as possible, measures to deal with each one. Yet, the sudden and violent death of a person that I was responsible for and had grown to love was not a scenario for which I could easily fortify myself, or those serving with me, against.

A common example of a personal dilemma when one is in a job that requires putting people in harm's way is that of a company commander who is required to clear lanes of possible IEDs so that his Marines can progress forward. His job is to call up his Marines who are armed with antipersonnel obstacle breaching systems (APOBSs) to clear the lanes. He does so. The fighting becomes heavy, and his gunners must expose themselves to fire to get into a good position. Word gets back to the commander that the Marines were killed. It was his job to make the choice to send them into danger. These are the tough decisions that leaders and individuals make nearly every second they are on the field of battle. The problem, in this specific situation, was that a film crew was with the commander at that moment in the middle of the battle, asking him the hard questions in real time. Leaders deal with their decisions, think about them, and attempt to justify them after the fact, when the fight is over. Leaders cannot allow themselves to focus on these questions in the middle of the kinetics of battle or more lives could be lost. Death on the battlefield is inevitable and heartbreaking and is also one of the greatest challenges to a leader's ability to stay above emotion and to continue to perform his or her duties.

Several months before I deployed to Marjah in December 2009, my wife wisely gave me a journal for Christmas. I was taking command of the battalion and was originally scheduled to deploy to Iraq shortly thereafter. On the inside cover, she inscribed the following:

Dearest Brian,

We will all miss you so much while you are away, but I thought you could use this journal to capture the moments that remind you why you do what you do.

The kids are always asking your reasons for choosing the Marine Corps. I thought it would be neat for you to share those special moments with them that speak to your heart and confirm your decision.

I love you beyond words. You make us proud. Be safe, my love. Big prayers.

Always and then some—Forever in my heart you are with me —everyday!

I love you.

Nik

This journal provided me more than just an opportunity to express why I do what I do; it also provided me with a way to work through those complicated times when sorrow and hurt met pride and honor.

One such time occurred in mid-February 2010. My battalion deployed as part of the surge into the Helmand Province in Afghanistan.[9] Specifically, the battalion was tasked with taking part in seizing Marjah, a well-known Taliban stronghold. In the early 1950s, the U.S. Agency for International

---

[9] In late 2009, the United States committed to a plan for a surge of military troops, the majority of which were Marines, to Afghanistan for a counterinsurgency strategy. See Kummer, *U.S. Marines in Afghanistan, 2001–2009*.

Development (USAID) was instrumental in developing this large farming community that included an extensive canal system. The Taliban later used it as a logistics hub and prominent poppy growing area. My unit initiated a combined assault on Marjah. The battalion was tasked with moving from north to south. Fighting was significant, and I spent a lot of time forward with the Marines and sailors in addition to the unit's Afghan partners as they assaulted through their objectives. They fought with courage, restraint, and effectiveness. One evening, the battalion suffered its first combat loss. I remember moving my jump platoon to the evacuation scene. I also recall being angry, not at the Marines but at myself. I said a prayer over the fallen Marine's body; I spoke to his squad members and reminded them and their company leadership that we had just begun the fight and that we would grieve later. I reminded them of the battalion's mission, of the importance of their efforts in making a difference, and to hold the line. I returned to our forward operating base (FOB), escorting the casualty to the BAS, and then headed to the command post where I received the battlefield update brief and sat with the intelligence officer.

Then, when time allowed, I settled in to write personal condolence letters to the Marine's pregnant wife and his father—another emotional punch with many more to follow. It hit me hard when I wrote the letters to moms, dads, and wives—giving notice to the living, the ones left behind, who may have to make explanations to children and babies not yet born. I wanted them to know how proud I would always be of their Marines and sailors, and that they should be proud too. I wanted them to understand. I wanted them to forgive. Because I put each family member's feelings in mind as I wrote, I wept during the writing of each letter. It hurt, but it was necessary. It allowed me to get through the emotion and provided the families details that I know I would want to know if it were my loved one. In fact, it was my loved one too. Putting pen to paper to pay respects to the family was the very least I could do to honor the memory of each fallen brother. When I was finished with the correspondence, on 16 February 2010, I wrote the following in my journal, recording the loss for myself:

It seems a bit off that I would be writing in this book shortly after the first combat loss. . . . [The Marine was] married, wife expecting. I wrote his father and his wife a letter, not an easy task. I spoke to his squad leader and members of his squad, not an easy task. I spoke to his young Lt—not an easy task and I spoke to GOD. Not always an easy task. In all of these I found strength, I gained confidence, [and] I reaffirmed my deep affection for my Marines and why I love being a part of this organization.

Shortly after the worst of the fighting, the unit arranged its first in-theater memorial service to celebrate the lives of four of our Marines, including our first loss. It was the opportunity, promised earlier, to grieve and say good-bye. We bonded, we said good-bye, and for an hour, the fight stopped for those in attendance, allowing them time to share their emotions. Then I had to draw it to a close; I had to return them and myself to the warrior's mind-set for our own sake and the sake of the mission. We had to close the ranks and continue to hold the line. These were my words to the Marines and sailors:

Good Afternoon,

I say "good" because today I have the privilege of honoring our fallen Marines and addressing their brothers, some of the finest Marines and sailors I have served with. This memorial is simply about two things. Honoring the ultimate sacrifice of [the four Marines]; and allowing their brothers to say goodbye until another day and do what they would want them to—which is to continue the mission. Soon you will hear from the Commanders and Marines closest to these heroes. They will speak of experiences and genuine feelings that create the unbreakable bond of brotherhood not truly understood until actually being a part of it. I only wish that we were able to share this time with the entire battalion as your emotions will collide. All that you have experienced so far will rush through you, overwhelming your senses, making you question your ability to continue—let the wave through—share your emotion, your love for your brother, your fear for the future, then realize as it passes that you are not alone, that because of the sacrifice of these warriors we honor today, the warriors around you today, and the warriors on patrol, on post and holding the line—you have nothing to fear—that this is the time to say goodbye to our brothers and to retake our positions in the line—whose strength depends on our commitment to the mission and each other and ensures that the ultimate sacrifice of these fallen warriors was not in vain. These warriors have left behind mothers and fathers, wives and unborn children and us, their brothers in arms—but I am here today to tell you they have not left us. They watch above us—feeling their presence gives us strength and purpose—and fond memories of the past brings smiles and laughter in our celebration of their life. They will always be with us—and we owe it to them to make it positive and celebratory. Honor them with exuberance, love, and understanding and do what they would do—and continue to hold the line. Each night before I rest, I pray these words: "Visit this place, O Lord, and drive from it all the snares of the enemy; let your holy angels dwell with us to preserve us in peace; and let your blessings be upon us always; through Jesus Christ our Lord, Amen."

(Names of the fallen Marines) are amongst those angels now—be glad in that.

As I concluded my commemoration, each company first sergeant conducted a roll call of his Marines and sailors. One by one, they emphasized the missing Marines by repeating their names three times with no answer. The company first sergeants' voices continue to echo in my head: Lance Corporal (last name) . . . Lance Corporal (first name, last name) . . . Lance Corporal (first name, middle name, last name)," and 21 shots were fired, "Taps" was played by a lone bugler, and I paid my respects along with my Marines and sailors. Some walked up together, others alone. Some kneeled, others stood. Some prayed. Others were silent. All of this happened in front of a pair of boots, a rifle, a helmet, and dangling dog tags with the background of our national ensign and the battalion battle colors.[10] The senior leaders, the sergeant major, and I were the first to pay respect to the fallen Marines.

Once we finished, the sergeant major and I stood to the side and, as the Marines and sailors filed through, we reinforced our love, our concern, and our commitment by simply giving them a hug and standing firm. There is not a day that goes by that I do not think about the 10 Marines, 10 Afghan soldiers, and the one interpreter I lost. They were my responsibility, and I loved them. Evacuating wounded warriors is hard enough, but watching my deceased go to their final rest was much worse. Several memorials were conducted throughout the deployment, and they took a toll over time. On 6 May 2010, I wrote the following:

Because he would want you to.

Because you have to.

You have to close the ranks,

And hold the line.

We will mourn another day.

Close the ranks.

Hold the line.

My words to the Platoon of Company I after putting (name) on the Angel Flight out of the LZ [landing zone] and following a prayer from the Chaplain.

I closed this journal entry with "They say these are the times that try men's souls . . . I will submit to you that my heart aches, my head throbs, my eyes water, my knees loosen, and my soul is tired."

War is hell. And then, it is over, but it is never really over. I will never forget all that we saw and experienced, but I will cope. I know it is part and parcel of the life every warrior has chosen. I know I put my heart and soul into preparing them; it is what we as Marines do. I ensured that my

---

[10] A national ensign refers to the American flag displayed onboard ships of the U.S. Navy and at Navy and Marine Corps commands on land.

subordinate leaders and myself were prepared to process our own grief and help the Marines and sailors navigate through their personal emotional quagmires. In the moment, we were not distracted by the logistics of executing the ceremony. The ceremonial details were practiced and mastered as a critical part of our predeployment readiness. The sergeant major, along with the company first sergeants, developed a plan, acquired the resources to professionally conduct these services, and prepared a letter of instruction (LOI) for everyone to follow.[11] This plan included the capability to record the ceremony and the means to deliver a copy of the recording to the family of each fallen servicemember. The loss of my Marines and sailors broke my heart, but for the sake of my unit members and their families, I maintained my professional bearing and stayed focused on the mission.

Preparation *is* the mother of victory. And yet, there was simply no adequate mechanism to ready myself for the time I would spend with the parents, spouses, and the children of the fallen Marines and sailors upon my unit's return from Afghanistan. I had written my feelings in letters, but eventually, I had to face the families. I had led their loved ones to war. I had trained them hard and prepared them for the worst so that I could bring them back safely, and ultimately, I failed. I could not change that harsh reality. Soon after the unit returned from Afghanistan, we conducted a battalion memorial service. We arranged a dinner the night before the service for family members so I could meet with them individually. That evening, I revisited all of the emotions that I had attempted to leave on the battlefield. I said a prayer, and with apprehension and fear, I approached each family. Some understood what I was trying to do and what I needed to do, and they hugged me and amazed me with their kindness and love. Others could not look at me and wanted nothing to do with me. I tried to understand, but it stung, and those feelings still stay with me as much as the memories I have of the Marines and unit members who I did not return to their families' embrace. I failed them, the families; what did I expect?

---

[11] The LOI provides the details for conducting a desired activity or action to include a mission, specified tasks, and coordinating instructions to ensure necessary items are not overlooked during the stress of the moment.

On the following morning, the battalion held the memorial service. The service allowed the families to have a chance to meet and grieve together; to know that they were not alone; to see and meet their loved ones' other family (battalion members); to realize that they were a part of that family; and to gain a sense of the pride, love, sorrow, grief, and loss that I, as the commander, and the unit felt so intensely. It was my solemn prayer and the deepest desire of my heart that this time of gathering together would convey what words could not, that the families would feel the fierce love and profound respect that those on the parade field felt for the fallen servicemembers. I prayed the families would find peace in knowing their loved one did not die in vain.

Furthermore, I understood that there were other heavy hearts that needed to be acknowledged during the service. I was not just addressing the families of the fallen. I was readdressing the emotions, the heartache, the guilt, and the sorrow of their extended families—their brothers-in-arms, who had returned and were revisiting their own emotions that had been necessarily suppressed during combat and successfully stowed away in order to close their ranks and fight another day. The difference now was the Marines and sailors were no longer in full-body armor, protecting the men around them. In the midst of dealing with the families' emotions, I also had the feelings of my Marines and sailors, as well as my own, to consider and to keep in check. It was a challenge unlike any other.

Finally, just as on the battlefield, in life, we must move forward. Now, the key is being there for one another, always. Regardless of the time of day or activity, we must be willing to pick up the phone or knock on the door. When the phone or doorbell rings, we must answer. We cope, and when it becomes too difficult for us to cope alone, we call or visit a fellow Marine or sailor who knows, who picks up the phone or answers the door. Someone who listens then closes the ranks to hold the line. War is certainly hell. Unfortunately, for those who serve, it is not over at the end of a deployment or with a cease-fire, treaty, victory, or defeat. The memories remain, but we always reform the line and move forward. That is what we do.

REFLECTION POINTS

- Considering the stresses faced by Marines, soldiers, and sailors, what role does leadership, as described by Lieutenant Colonel Christmas, play in addressing the emotional needs of servicemembers deployed to war? Does acknowledging emotion make it easier for these men and women to face trauma and to be resilient?

- What can military leaders do to prepare families for deployments? Are the institutional forms of support or the personal touch of the commanders more important? Or, do the resources work well together in providing assistance to families during wartime?

- While battalion and unit leaders must train their warriors for battle, what should they consider when preparing families? Do these two aspects of predeployment activities conflict or complement each other?

- What can commanders do to prepare themselves for the possibility of casualties and deaths in their units during deployments? What resources are available?

- Considering the challenges of preparing a unit for deployment, how does a commander balance the need for training with the need to ensure family time? What is the proper ratio? How does a commander obtain support from higher commands, unit members, and families for a commander's assessment of the unit's readiness and what it needs to succeed during a wartime deployment?

- Discuss methods of how a servicemember can prepare and ultimately cope with combat losses both as a leader and a brother-in-arms.

CHAPTER THREE

A FELLOW MARINE IN COMBAT: MY BROTHER
CAPTAIN MATTHEW C. FALLON

Captain Matthew C. Fallon deployed two times as a first lieutenant with 3d Battalion, 9th Marines, first to Iraq in 2009 and then to Afghanistan in 2010–11. As an infantry officer, Fallon served at company and battalion levels. His story covers events during his deployment to Afghanistan. Fallon's brother, then-Second Lieutenant Timothy E. "Tim" Fallon, also an infantry officer, was wounded just before Fallon deployed. Then-Lieutenant Matthew Fallon was forced to leave his wife, new son, and extended family as they dealt with the trauma of and struggle to overcome the reality of his brother's wounds. Fallon's story extends through his own homecoming and reintegration with his family.

I am the oldest of three brothers. We were very close growing up. Tim was the second oldest and Dan the youngest. We spent most of our days playing in the woods around the house reenacting the adventure stories my mother would read to us. Throughout high school and college, we remained close. Even in college, Tim and I frequently saw each other and hung out. We could not have been closer.

My uncle was a Marine infantry officer, one grandfather had been in the U.S. Army 82d Airborne Division, and my other grandfather served in the U.S. Navy onboard a destroyer escort in the Battle of the Atlantic during World War II. None of them shared any serious war stories with the exception of my uncle after I started to show an interest in joining the Marine Corps. He told me several stories about combat in Operations Desert Shield/Storm and a few about Panama. As I matured and read more adventure and war stories, I developed a fascination with military history, which grew into a desire to serve.

I received a Naval Reserve Officer Training Corps (NROTC) scholarship to Villanova University in 2003. I can still remember the selection officer giving the speech at my high school. I was to be part of the "tip of the spear," the leading edge of America's military. After graduating Officer Candidate School (OCS) in 2006 and college in 2007, I went to The Basic School followed by the Infantry Officer Course (IOC). During that time, Tim went the other direction, embracing college life to the extreme; he grew a huge red beard and long curly hair. Though he had many interests similar to mine, I never expected him to join the Marine Corps.

After graduating IOC in 2008, I was assigned to 3d Battalion, 9th Marines, a newly activated battalion at the time. I became the M252 81mm mortar platoon commander. We trained constantly trying to weld the new unit. I was married on 27 September 2008 to my college sweetheart, Jen, and less than a month later, I left for training. And while the next few months required more training locally, I was gone for most of the week. Jen hardly ever saw me. I distinctly recall coming home one day to meet a dog my wife had brought home from a shelter. Training continued into the spring and focused on improving the ability to rapidly and accurately shoot mortars. The unit was preparing to go to Afghanistan. In April 2009, we learned we were going to Iraq instead.

I was tremendously let down by the news. My platoon and I had worked furiously to prepare, and now we were going to a war that was over.[1] We arrived in June 2009. I went "out of the wire," outside the American-controlled safe perimeter, only once while in Iraq. At that point, U.S. troops had to receive permission to enter the cities. Most of our time was spent drilling as the heliborne quick reaction force (QRF) for all of al-Anbar Province. If something catastrophic occurred or an aircraft went down, we launched to secure the site. Only a commanding general could launch the QRF. My primary challenge was to keep the Marines under my command focused, and when we found out we were leaving early, that became nearly impossible.

After an uneventful tour in Iraq, we returned to MCB Camp Lejeune and immediately prepared for potential redeployment as part of the opening stages of the Afghanistan troop surge. We were told to prepare to participate in the clearing of Marjah, a rural community and Taliban stronghold in the Helmand Province of Afghanistan. By December 2009, it became clear that we were not one of the battalions that would deploy. The 3d Battalion began to shed Marines, who left the Service or went on to other jobs in the Marine Corps, since we were no longer on a deployment footing. Within the battalion, many people switched jobs or were promoted to make room for incoming Marines. I left the 81mm platoon and became the battalion assistant operations officer.

Meanwhile my brother, Tim, graduated from college early and decided to become a Marine officer. I was in shock, but could not be more proud. He braved a winter OCS class, then continued straight to TBS and on to IOC. He became a platoon commander in a sister battalion, the 2d Battalion, 9th Marines. He lived 40 minutes from the base on "Lieutenant Isle," and so he became a part-time resident at my house since I lived on Lejeune, only five minutes from where he worked, giving him the opportunity to bring pickles to my wife during her pregnancy. Tim and I spent long nights debating tactics and training. We became even closer in those few months. He deployed first, and it was hard to see him go as I waited for my deployment to begin.

---

[1] The U.S.-Iraq Status of Forces Agreement signed by President George W. Bush in 2008 called for the withdrawal of U.S. combat troops from Iraqi cities by 30 June 2009. In February 2009, President Barack H. Obama extended the combat withdrawal date to 31 August 2010.

Tim left for Marjah in August 2010. I saw him off as his company boarded buses. There were hundreds of Marines waiting for the buses and their families saying goodbye. I am embarrassed to say but, it was the first time that the weight of being a Marine leader really struck me. Fathers said goodbye to their children and tried to explain—for the thousandth time—what seven months away meant. Mothers cried as they said goodbye to their Marine sons. Girlfriends sobbed. Some struggled to remain emotionless. Others obviously just wanted their loved ones on the bus and away to escape the depth of emotion. The terrible weight of responsibility for another's loved one became deeply apparent as I said my own farewell to Tim. The worst moment was watching the Marines board the buses. Tim was nervous and, at the same time, frustrated with the long process. I distinctly remember thinking that some of these men would not come home; some would not come home in one piece. I do not know why, but I had not truly grasped the solemn human element of deployment before this. My deployment to Iraq, so late in that war, was just not dangerous enough to bring this realization home, especially after preparing for clearing operations in Afghanistan.

In the weeks following my brother's deployment, Tim and I occasionally talked on the phone. He was doing what he had trained for and was enjoying it. I was enjoying my job as the assistant operations officer. My role was mostly planning and supervising the execution of the details of the unit's plans. My unit, 3d Battalion, 9th Marines, was scheduled to deploy around mid-November to Marjah proper. Tim and I joked that we would see each other in-country. At the same time, my wife was very pregnant. With a little less than a month before I was to deploy, my son Emmett was born. We had a christening party with our families and looked forward to the following summer when we would all be reunited. My youngest brother Dan was, by this time, a midshipman in the NROTC unit at George Washington University.

One night, a few days after my family had left, Emmett would not sleep. He was bawling. Car rides calmed him down, so I strapped him in the baby seat and drove around Camp Lejeune. My father called me on the way home: "Where are you?" Reporting on my fatherly duty, I replied, "I'm driving around trying to put Emmett to sleep. Why?" There was a rare urgency in his voice, and he said, "Pull over." In that instant, I knew Tim had been hurt. For a few seconds, I thought he was dead. I pulled over. He told me what they knew. It was not much.

The initial report was that Tim was wounded in the face. What did that mean? Well, the official could not tell us because he only read from a vague and limited report. Tim and I had talked over beers about what part of the body would be better to lose and agreed that losing a leg was not so bad. Being wounded in the face had never entered the conversation. Was he disfigured? Brain-dead? Blind? No one could tell us. Despite the immediacy of social media and the Internet, it remains difficult to talk to those who are in the fight, and information that does come out is often inaccurate. It is agonizing for family members as they desperately try to piece together shards of accurate information to create a picture of the extent of their loved one's injuries. It will always be that way.

I drove home and told my wife. She was in shock. I watched a YouTube video with particularly powerful music again and again. It seemed to vaguely reflect my emotions but could not match the depth of what I felt. Eventually, I went to the office and e-mailed Tim's company commander, who I had met once. I hoped that I would hear from him by the next morning. My wife and I went to bed feeling empty and incomplete.

The next morning, I went into work overwhelmed by what had happened and concerned about what the future held for both my brother and my family. I was numb, moving in slow motion, yet I was scheduled to deploy to Marjah in two days. I remember seeing my battalion commander Lieutenant Colonel David W. Hudspeth who asked, "How's it going, Matt?" I replied, remaining stoic, "Tim's been hit." The commander's response, "I know I just saw the casualty list," was full of meaning, yet there was no other way for me to react but unemotionally. The fact that he knew and was obviously seeking me out to talk about it made me feel better. I heard from Tim's company commander, but he had very little information as to the extent of the injuries. In combat, we do not treat and diagnose the wounded. We just stabilize and evacuate the injured. The hospital units in the rear areas are staffed with true professionals who diagnose and provide treatment. Despite twenty-first century communication technology, there is still a "fog of war" at the front where the wounded are rapidly triaged and transported.

My family anxiously sought more information, so my uncle, who knew Tim's battalion commander, sent him an e-mail. The battalion commander said that Tim was just fine and would recover with only superficial damage to his face. I remember thinking this was too good to be true. So, despite the good news, I continued to hunt down information. I talked to the medical chief and got the number to a hospital unit in Afghanistan. After hours of phone calls, I tracked down a sergeant who was in charge of routing casualties to the United States. She read Tim's chart to me, which was against regulations but merciful. His face was structurally intact, but both eyes were badly damaged. I did not know about brain function or much of the medical terminology used by the sergeant, so I could only understand pieces of the jargon. I called my mother and painfully delivered the news that Tim may be blind and have a traumatic brain injury (TBI). Although my family and I were never speechless, our words never captured the depth of our distress. I remember conversations as being surreal, skimming the surface of our agony as each person tried to be strong for the other. If one person fell apart, we would all fall apart. My mother called my father and Dan to give our family the news of Tim's injuries. Yet, there was still Tim's fiancée who needed to be notified, and I had to make that call. There was not much to say except to repeat the little we knew. But the little we knew was weighty enough, and the unknown was even more solemn. My unit's deployment was delayed by 48 hours. As I prepared to deploy to Afghanistan in the next two days, Tim was on his way to the states to Walter Reed Army Medical Center in Washington, DC.

Through a friend in the U.S. Air Force, I was able to get Tim's flight information to the United States. If I had not had that contact, we never would have known when Tim arrived in the states. He was scheduled to arrive 36 hours before I was scheduled to leave. I considered traveling to Walter Reed where I might have been able to see him, but my mother talked me out of it, urging me to spend every moment before I left with Jen and Emmett. So along with the rest of my unit's advanced party, I packed my kit and staged to load the buses. I recall waiting to leave. A friend asked, "Isn't your brother there?" meaning Afghanistan, of course. I responded simply, "No, he was stuck by an IED." That was the first time that most of my friends heard of Tim's injury.

I relied on emotional distance at this point. My uncle called and told me they, the family, had Tim, and I had to concentrate on what I had to do. This was repeated by my father and mother. In retrospect, it is remarkable to even realize that only 96 hours transpired between my learning that Tim was hit and my leaving for Marjah. I was in an emotional haze of anger. Leaving my wife and 12-day-old son hardly registered. I had not even had the time to comprehend, beyond the intellectual level, that I was a father. As I got on the bus, a profound moment occurred that I will never forget. For unknown reasons, the buses were delayed multiple times. The "Old Man" (the commanding officer of the battalion) had to get off twice to explain to his little girls that he was leaving. He came back in a rage: "Don't these motherf——s know how hard it is to explain eight months? Let's get f——g going." A few minutes later the Old Man made a joke, and I think he felt bad about destroying us, but it was for a good reason. Then he said, "You know, Matt, you're gonna be a better officer now that you're a father." I said something shallow but polite in response. He responded, "No, you don't understand. John A. Lejeune said to love your men like you love your own children. But you're gonna find that no matter how hard you try, you will fail at this, so you are constantly going to struggle to love them [your Marines] the way you have been charged to." It was among the best leadership advice I ever received, made more powerful by the timing of its delivery.

The 3d Battalion, 9th Marines, was part of the same regimental combat team that the 2d Battalion, 9th Marines, (Tim's battalion) belonged to, and I was deployed only three miles away from what had been Tim's outpost in Afghanistan. People looked at my name tape then at me and said, "Are you the Lieutenant Fallon that was wounded? You can't be." I answered truthfully that it was my brother. In retrospect, I sometimes wished I had just said, "No." I tried to keep the focus on what I was there to do, but that was difficult because of the circumstances. I did a lot of planning work to bring the battalion into the country and deploy to Afghanistan. There is a running joke in the Marine Corps that the two worst units in the Corps are the one you replace and the one that replaces you. This was not true of 2d Battalion, 6th Marines, the unit that we replaced. We listened to what the 2d Battalion, 6th Marines, had to say because that unit had fought hard and well and shattered the Taliban in Marjah. Marines from the 3d Battalion, 9th Marines, had a lot to learn. First, my unit had to understand how the Marines had fought and thought in this particular place. Much of the time was spent talking through what had happened before the 3d Battalion, 9th Marines, got there. We had to learn about our Afghan partners—both the official Afghan National Security Force (ANSF) and local security forces. Additionally, we had significant planning to coordinate with higher headquarters. One constant surprise was the amount of gear the unit received from overseas that we had never trained with or heard of. Some of it was highly technical and complex, while other pieces were as simple as a scythe on a stick used to find wires or the soft spot in the earth where potentially dangerous items were buried. Throughout the early months of my deployment, my family told me that Tim wanted to talk to me. Although phone calls were successfully arranged between Walter Reed Army Medical Center and Marjah, I found that I almost did not want to talk too much. I was not sure what to say to Tim, and I knew I had to focus on where I was and what I had to accomplish. Again, I kept myself purposefully distant to maintain emotional control.

Then came the first operation, and I was tasked with planning it. It was a reinforced company clearing mission to deny the enemy an assembly area near the heart of the town. I went on a reconnaissance patrol with the engineer platoon and found the key problem: we had to resupply the company and push down enough engineering supplies to establish a patrol base. The battalion was worried, during this time, about the prospect of a welcoming party, when the enemy tests a newly deployed battalion by attacking everywhere. Even a week or two prior to turning over the battle space to the 3d Battalion, 9th Marines, every single position held by the 2d Battalion, 6th Marines, had been attacked within a few hours. Something similar never occurred during my unit's deployment. We later determined that a deception plan we had run was likely the cause of the enemy's hesitancy to act. I remained in the battalion combat operations center (COC) during the first clearing operation. The boring nature of this work is vile. For what seems to be an eternity but can only be minutes, one sits waiting for a report. But for Marines at higher headquarters, they shut up and stay off the radios because the "dudes"—the riflemen and small unit leaders in the field, those at the cutting edge— know what is going on and what to do. The long minutes are followed by brief moments of stress and excitement from news of cache finds or clearing detonations and clearing fires or giving directions to a forward element. That is followed by more waiting while the Marines in the field do their work. It is also immensely frustrating since someone in my position could not be there, and all I wanted was to be there with the Marines.

During officer training, instructors constantly talked about the bond we officers would form with our first Marines and how much counseling would be required in our roles as leaders. The first Marine I ever counseled was from a small town in Louisiana. He had graduated infantry school the week before and was about to marry his high school sweetheart. He was 18 or 19 at the time. He remarked that the roads around Camp Lejeune were the biggest he had ever seen. It is very difficult for people so young to be married and more difficult to be in a place so far from home and so alien in culture. Both he and his fiancée knew very few people outside of their hometown in Louisiana, and they were about to move into Jacksonville, North Carolina, which, while a modest town, was much larger than his hometown. After I was no longer his platoon commander, I learned that he and his wife were experiencing the typical stresses of a young couple.[2] When word came that this Marine was wounded, it was a blow. I had known him and liked him as a Marine, and while he was not picture perfect, he worked hard and persevered no matter the situation. Then, he became a double amputee. Later, when reports came in from the hospital (reports from units in the fight are by necessity short and only contain the most necessary details), I learned his penis had been destroyed.[3] I remember the depths of my sadness; the tragedy of someone that young dealing with the normal challenges of newly married life and coming home mangled.

For many servicemen and servicewomen during the Iraq and Afghanistan wars, the daily grind of combat was more about drafting reports and e-mails to request support assets and watching intelligence, surveillance, and reconnaissance (ISR) feeds. At the battalion level, one knows the people fighting, but above that, I found a tendency for people to lose the personal element of war and become consumed with the requirements of daily duties. From the grunt's perspective, it was very hard to watch Marines be in the fight without going into the field alongside them. Occasionally, I would leave my primary role in the operations center to go on patrols through the area of operations. There was a single time I was actually in a significant fight. One company was pushing into an area east of Marjah, controlled either by the Taliban or a drug lord. It was a reconnaissance in force operation to determine routes and the location of a new patrol base in

---

[2] The Marine and Family Programs Division, under the deputy commandant for Manpower and Reserve Affairs, is tasked with providing policy and resources to support personal and family readiness programs. One such program, LifeSkills Training and Education, addresses family functioning and communication for Marines and their families. See the Marine Corps Community Services website at www.usmc-mccs.org.

[3] Injury to the genitals, bladder, urinary tract, and kidneys, known as genitourinary trauma, became more common during the wars in Iraq and Afghanistan as American troops participated in dismounted patrols with frequent exposure to IEDs. See Sherrie L. Wilcox, Ashley Schuyler, and Anthony M. Hassan, *Genitourniary Trauma in the Military: Impact, Prevention, and Recommendations*, CIR Policy Brief March 2015 (Los Angeles, CA: University of Southern California, Center for Innovation and Research on Veterans and Military Families, 2015); and Dismounted Complex Blast Injury Task Force, *Dismounted Complex Blast Injury: Report of the Army Dismounted Complex Blast Injury Task Force* (Fort Sam Houston, TX: Army Surgeon General, 2011).

support of a regimental push within the month.[4] I knew the Marines were going to see enemy action, because every time we moved east, the enemy would invariably provoke a fight. Knowing this, I went with the Protective Security Detachment (PSD) trucks to see the place where we would be operating for the next few months. I clearly remember prepping gear the night prior and the morning of because a dry lightning storm eerily created spider web designs through the low clouds. In movies, occurrences such as this are portrayed as omens, but while on the ground during war, they are often just passing observations. The trucks moved to one of the patrol bases on the edge of what was, in essence, the front line. If we crossed into Trek Nawa, the enemy always came out to fight.[5] While we waited for the cloud cover and rain to clear, the company commander reviewed the plan to the battalion commander. The expectation of getting into a fight was foremost on my mind. By midmorning, the rain stopped, and the company began to move out of patrol bases into Trek Nawa. Much of the first hour was consumed with pushing through fields. We followed the exact path marked in front of us. Up front, the sweepers led the way with metal detectors and scythes. I am loath to admit that, like many Marines new to combat, I longed to be in a fight, and part way through the push when nothing was happening, I began to worry that we would not encounter action.

Just as it seemed that this would be an uneventful operation, there was the crack of rifle fire. It was only a few shots at first. The formations spread out and began to push across the fields while I stopped and held security for one of the designated marksman (DM) with the PSD. He was providing overwatch as the rest of the company moved forward.[6] I could not see anyone, nor could I really fire as columns of squads pushed in front of me. Minutes went by, and it dawned on me that we were all alone—two men on security and the designated marksman, who was on the roof of the compound. The entire force was 200–300 meters ahead. "Baily," I shouted to the

---

[4] The term *reconnaissance in force* refers to an operation designed to gather intelligence about an enemy's capabilities.

[5] Trek Nawa is located east of Marjah in the Helmand Province.

[6] The term *overwatch* means to provide fire support for another moving element.

DM, "I think it's time to go." Everything comes out like a cliché in combat. He jumped off the roof, and the three of us sprinted across a field. "This is stupid," I thought. IEDs are rarely in the middle of a field, but rounds were cracking overhead. Suddenly, I had an image in my mind of multiple YouTube helmet-camera videos portraying this moment. We ran 200 meters; mud from the rain started to clump on our boots, slowing us up. I was in front, moving across the field. I sensed the other two had fallen back but were still with me.

After what seemed like an eternity, we reached a compound on the far side of the field with the rest of the PSD. After a few minutes, we moved forward to one of the outbuildings farther south by about 50–60 meters. Somehow, we ended up on the left flank of the company's advance. We started to take a higher volume of fire, and for some foolish reason, I moved out from behind the building. I wanted to get to a strong firing position to help suppress the enemy team or squad that had us pinned. I sprinted from behind the building back to one of the walls 10–15 meters away. I fired my weapon for the first time in anger at one of the possible firing positions. I distinctly recall watching my weapon's plastic barrel protector fly into pieces as the round flew out. Firing the first round always takes more oomph. Once outbound, subsequent rounds come much faster. Some Marines from the company came past me and continued south toward the village, which was our objective. Increasingly, I began to realize my position was imprudent. If I had been shot, no one would have seen me. Having come to this earth-shattering conclusion, I moved back to the building next to everyone else. At this point, we were pinned down. When we tried to move from behind the building, there was a crack from gunfire. To gain fire superiority, the DM climbed up onto the roof, and within seconds, he and his rifle came tumbling back off as all of the enemy fire concentrated on him.

At this point, gun trucks came forward and put enemy positions under fire. We were able to achieve fire superiority. We bounded forward in two groups, one providing fire for the other. This maneuver was more organized than the first run across the field. We bounded between 50 meters and 100 meters twice. It was the second bound forward that the combined antiarmor team's (CAAT's) grenades from the automatic grenade launcher began to fall short. The grenades were landing within 50 meters of the fighting Marines, and it took a few minutes to shut off. The Bell AH-1 Super Cobra helicopters finally showed up.[7] The relief of witnessing an enemy position pulverized by rockets and 20mm cannon fire is hard to describe. After one or two more runs, the enemy stopped fighting; rarely would they fight with aircraft on station such as Cobras or certain drones. With the firepower to suppress enemy fire, the company cleared to the objective; engineers began conducting their reconnaissance of the compound that would be the patrol base, determining what other supplies would be required to turn it into a permanent position for Marines.

There were only a few people on security. Snipers established an overwatch position to allow us to safely perform our duties. As the engineers worked, the cloud cover from the morning came back, which was a serious problem because aircraft would not be able to fly. We could see movement to the east. Just as the engineers finished their reconnaissance of the future patrol base, the enemy counterattacked by fire. One of the platoons suppressed the enemy while the rest of the company began to withdraw as we had the information required for future operations in the area. It took more than an hour to walk across the muddy fields back to the established patrol bases. It was just a matter of each Marine following the guy in front of him to avoid the IEDs.

Three weeks later, the actual clearing operation began. The goal was to seize the staging areas that anti-Coalition forces were using in Trek Nawa, east of Marjah. To do this the battalion would establish a patrol base in the position that engineers had found earlier. This time I remained in the main operations center. The forward operations center, consisting of trucks specially outfitted with communications equipment, had moved forward to control the operations in Trek Nawa. The first few hours of the operation went well. There were close to 800 Marines, members of the ANSF, and local

---

[7] The Bell AH-1W is officially named the Sea Cobra, which is virtually never used. It is almost universally known as the SuperCobra or Super Cobra, but will be referred to throughout this book as the Cobra.

fighters called Interim Security for Critical Infrastructure (ISCI). With such a large number of people pushing into the area, Taliban fighters and foot soldiers for the drug lords were unlikely to take the opportunity to fight us. The odds were too much in our favor. However, around midmorning, a call came from the units on the northern flanks. There had been an IED strike but no casualties. When the unmanned aerial vehicle (UAV) finally positioned overhead, we saw that the mine-resistant ambush protected (MRAP) all-terrain vehicle (MATV) was torn in half. It was impossible that there were no casualties. After several minutes, a corrected report finally came through.

The nature of war is always complex and often uncertain as so many men and women try to carry out their isolated tasks while coordinating with others in a naturally dynamic and chaotic environment. No one will ever know how that initial, very wrong, causality report reached the main COC. It did not matter; those who died would have died even if accurate reports had been received. This type of misinformation is a common occurrence on the battlefield. The Marines on the deck pulled the five casualties out of the truck and recovered one Marine from a canal 20 meters away. That Marine was the gunner who had been ejected from the vehicle by the force of the explosion. Two Marines died soon after the evacuation helicopter took off. The other three were marred for life. All of them were from Weapons Company, and I knew several of them, including the vehicle commander who died. He was a good man. He had risen from a private first class to a sergeant in three years and deserved every bit of it. Everyone respected him as a level-headed and fair leader, who was always respectful but never intimidated by rank. He was the heart and soul of that section.

When the vehicle was recovered and brought back to the battalion's forward operating base, the MATV's condition was jarring. People who work on large bases and camps are cut off from the daily grind of fighting and can lose touch with the brutality of war. There was significant shock among the Marines as they stared at the remains of the vehicle. Most Marines at the FOB gathered around, mouths agape. The vehicle had been literally ripped apart. It was horrific and terrifying to see it torn to shreds when it was supposed to provide so much protection. I wonder if after the landings in Normandy, troops looked at those M4 Sherman battle tanks, gutted by the German Tiger heavy tanks and 88mm guns, and felt equally terrified.

While the unit's operation was successful, the presence of those IEDs forced us to fight much longer in Trek Nawa than we had intended. Two months later, the battalion started to redeploy. I was among the first out. Throughout the entire deployment, I had rarely talked to my family, my wife, or Tim. Admittedly, this was a half-conscious decision. I did not want to think about the trauma at home. It was as if I could not simultaneously handle fighting a war while coming to terms with Tim's wounds. Throughout this time, we had found out that Tim would not be able to see again without tremendous technological advancements. He and his fiancée committed to stay together; they became a unit of strength. There had been things I heard about while I was gone, like the Christmas Eve Mass at our hometown church, St. Luke's. We had been parishioners since we were kids, and when Father Mike welcomed Tim home, there was a standing ovation and people lined up to shake his hand. While emotionally gratifying to hear these stories, because I knew Tim was being honored and supported by community, the stories were equally draining to hear and difficult to digest. I felt helpless when it came to Tim and his life-changing injury and that, in part, made me angry. But with my job in Afghanistan over, I was headed home.

Most Marines underestimate the enormity of coming back, probably because of the intense immediacy of a war's daily focus or the obvious traumas of combat. The majority of the men in the battalions had wives who had been running everything at home while the Marines were deployed. The wives single-handedly raised children, worked, dealt with finances, and made family decisions without input from their husbands. So no matter what a Marine has experienced, coming back after six months or a year, he has to face an intense process of acclimation. Very often his wife cannot understand what the Marine was tasked to do. Likewise, the Marine cannot understand what his wife went through. While I was deployed, my wife went through a period of postpartum depression alone. It was exacerbated by my own struggles to handle Tim's injury that made me unsupportive to her before I deployed. While I was away, I abruptly left her with a two-week-old child and headed to Afghanistan and turned inward after Tim's injury to focus on my job. To her credit, she did not share her burdens with me, but bore them herself to ensure I could focus on combat.

So, I faced a new challenge upon coming home, and it was one that I did not recognize until long after we had overcome the trials of war and returning home. First, there was dealing with Tim. While deployed, I had intellectualized his injury and his disability. I had separated myself from his pain. But the first time I embraced Tim after getting off the bus at Lejeune, I knew he had retained his trademark humor and love of life. Even a few months after his injury, he was running sightless down a dock and jumping into the White Oak River in North Carolina. However, to see the visual world now off limits to him was painful for me and triggered a raw ache that I had not anticipated. I experienced anger and frustration and had the sense that the world had moved off its axis because it had not properly recognized Tim's sacrifice.

Second, I was dealing with being a father for the first time, and diapers were the least of the trials. Finally, my wife and I faced the challenge of becoming a couple again after enduring emotional and physical separation and disconnection. There were months of postdeployment stress for both of us. My wife and I spent months circling each other and fighting for the sake of fighting. This was uncharacteristic of us as a couple, and neither of us really understood the reasons for the tension with one another. It was only after months of working together that we became strong friends again and were able to talk through the hurdles life had placed in our paths.

As for Tim, he has progressed faster than anyone could have anticipated. It is a testament to his resolve, his strength of character, and his love of life. While I was away, Tim had traveled around the country receiving superb care and rehabilitation at Walter Reed Army Medical Center with ophthalmologist Dr. Dal Chun, the Edward Jr. VA Hospital Central Blind Rehabilitation Center just outside of Chicago, and The Seeing Eye, Inc. in New

Jersey. He was fully supported by the Semper Fi Fund, Blinded Veterans Association, and Fisher House Foundation. Today, nearly six year later, he has a guide dog, a master's degree from Georgetown University, and a job and loves an 18-month-old son. I no longer have those days of anger when I only think about what a good officer he was and his potential, and bitterly wonder if Americans recognize his sacrifices for their safety. Yet, Tim's story reflects the experiences of thousands of Marines throughout history. Tim and I have talked, and we know our story is not unique. Other families have suffered far more in this war and in past wars. His life now embodies the strength of the human spirit when faced with great loss.

Tim first held my infant son, Emmett, 24 hours after arriving at Walter Reed Army Medical Center. Later, Tim bought Emmett a copy of J. R. R. Tolkien's *The Lord of the Rings*. In it, Tim wrote an inscription about the unique bond and love that develops between those who have served:

> I fought dark men far from home. I lost my eyes in battle against evil and the protection of that home. I would lose them again a hundred times over for you and the light my brothers and I protect. I never saw more clearly than when I held you for the first time in a hospital room, lightless only to me.

REFLECTION POINTS

- How do injuries of servicemen and -women affect leaders and their actions? Is there a way to prepare for these moments?

- How does poor communication or lack of notification affect those on the battlefield and stateside? If you have had experience with this, did it help not to get the worst of the news at once?

- Captain Fallon mentions uncertainty on several levels—the battlefield, the home front, and even his own feelings. How can policy makers, family members, and military leaders help Marines, soldiers, and sailors handle this uncertainty so that servicemembers can focus on wartime objectives?

CHAPTER FOUR

THE COBRA, THE CONVOY, AND A CRISIS OF FAITH
LIEUTENANT COLONEL WAYNE R. "GONZO" BEYER JR. (RET)

Lieutenant Colonel Wayne R. "Gonzo" Beyer Jr. served with the U.S. Marine Corps for more than 20 years as an aviation instructor and helicopter pilot of Bell AH-1W Super Cobras. He flew with Marine Light Attack Helicopter Squadron 169 (HMLA-169) and later with Marine Light Attack Helicopter Training Squadron 303 (HMLAT-303). During his service, he deployed twice from Okinawa, Japan, to Kuwait/Iraq for Operation Iraqi Freedom I (OIF I) and to Djibouti, Africa, for Combined Joint Task Force-Horn of Africa. Beyer narrates events that took place during his time as a captain deployed for OIF I in 2003 with HMLA-169. He was awarded nine Strike/Flight Air Medals and two Individual Action Air Medals with the combat "V" distinguishing device.

It was dark, there was no moon, and that meant there was very little contrast in the desert terrain for the night vision goggles (NVGs) to pick up. This made flying at low altitudes rather disorienting. It was 28 March 2003, and the I Marine Expeditionary Force (I MEF) was moving rapidly north from Kuwait toward Baghdad, Iraq, and I was flying in support from my

AH-1W Cobra attack helicopter. At this moment, my spatial world came down to two sources: the tracer fire from the Marine light armored vehicles (LAVs) blasting east into the buildings along the road and ricocheting high into the sky, and even heavier tracer fire blasting from the buildings to the west into the LAVs and ricocheting into the sky. The volume of fire was so heavy and in such a compact area that it looked like a scene from *Star Wars*. This was going to be the closest I had ever fired rockets to support Marines on the ground less than 100 meters away. To ensure we did not frag the Marines, my division had to push in really close, and that meant diving the Cobras right into that mess.[1] I was convinced there was no way we were not going to get hit by all those ricocheting rounds, but we had to stop those machine guns from unloading on the Marines.

> They said that without the Marine Close Air Support the losses to the convoy would have been significant.
>
> ~*Lieutenant Colonel Freddie J. Blish*[2]

How did I get here? It is a long story. Everyone has a unique path to becoming a Marine; mine began with the first Marine I ever met, my childhood priest Father Eugene Lutz. Father Lutz was the greatest priest I have ever known. He enlisted in the Marines at age 17, went to boot camp, and shipped out to an island in the Pacific. Unfortunately, that island was Bataan in the Philippines, and within a month of becoming a Marine, he found himself a Japanese prisoner of war and on the infamous death march. Story has it that while being tortured, he swore to God that if he got out of there alive, he would dedicate his life to God's service. He stayed true to his promise and became a priest. He was beloved by his community. Every now and then, he would wear a tattered USMC shirt. I could tell he was still proud of having been a Marine.

I learned early on that Marines are riflemen first. Officer Candidate School (OCS) and The Basic School (TBS) also engrained in me a theoretical

---

[1] Frag, in this case, refers to the killing of a Marine by a fellow Marine.
[2] LtCol Freddie J. Blish, e-mail to author, 4 August 2004.

appreciation for the Marine air-ground team. As a light and fast force, Marine infantry does not carry with it the same heavy firepower (e.g., tanks, artillery, heavy equipment, etc.) as the U.S. Army. Marine infantrymen rely on speed, surprise, and violence of action. They make up for the lack of heavy firepower with Marine aviation, which the Marine Corps has integrated into most aspects of operations from Marine assault pilots—who are legendary for carrying Marines into and out of hot landing zones—to close air support (CAS), where rotary-wing and fixed-wing aircraft integrate with the Marines on the ground to drop or fire ordnance in close proximity of those troops. If Marines are in contact with the enemy, Marine aviators are almost always there to supply them, protect them, or to get them out. And so, at TBS, a contract begins to be established between the aviators and the grunts. It is one of shared understanding, respect, and most importantly trust. Psychologically, it is brilliant to make all Marine officers go through the six-month training at TBS. There, we gain the mentality that we are Marine riflemen first and aviators second; the concept of "one team, one fight" is anchored there.

Between primary and intermediate flight training, I had gone home to ask Father Lutz if he would perform my marriage ceremony to my fiancée, Elizabeth. We had a long talk about marriage, and the consequences of becoming a Marine. He seemed torn—proud, but pensive about what might lay ahead for me. After all, his Marine Corps experience was more horror story than fairy tale. I confessed to him about having a conflict as well. In the grand scheme of life and faith, I was struggling with one concept I could not get out of my head. If we were called to aspire to be Christ-like, how could I reconcile that with volunteering to become a Marine aviator? I told him that I believe the Marines are ultimately a force for good, but I did not want to have to kill anyone. I realize there was an unfortunate reality in this world, and that someone has to be willing to stand up and fight evil. But I did not think Jesus Christ would choose this career path. Father Lutz put his hands on my shoulders, looked me straight in the eyes, and said, "That's why you need to become a Marine aviator. That kind of power needs to be in the discerning hands of someone who doesn't desire to use it." With tears in his eyes, he blessed my forehead and said, "God will be with you." It was the last time I saw him. He did not live long enough to perform the ceremony for us. Thousands came to his vigil.

During flight training, there was a lot of talk between my fellow Marines about what kind of platform (aircraft) we would end up flying, provided we were to make it through flight school. Prior to getting to flight school, most of us had an idea of what we wanted to fly. A lot of aviators wanted to select "jets," so they could fly McDonnell Douglas F/A-18 Hornets or McDonnell Douglas AV-8B Harriers. Others wanted to fly helicopters, so they could be closer to the Marines. As for me, I just wanted to finish flight school. At that time, they could give me anything, and as long as I got my wings, I was going to be happy. I did envision myself flying every platform, wondering which one would be the most fun to fly. I considered every platform, that is, except one: the Cobra. I am not sure why, but I discounted any notion of flying a Cobra. Even after primary flight training, when I was selected for the helicopter pipeline, I thought about all the helicopters except the Cobra.

The Marine Corps helicopter options at that point were the Sikorsky CH-53 Super/Sea Stallion heavy-lift helicopter, the Boeing Vertol CH-46 Sea Knight medium-lift helicopter, the Bell-Textron UH-1N Iroquois light-lift and utility helicopter, and the AH-1W Cobra attack helicopter.[3] Unless a pilot was at the top of the list of graduates in their class, it was a roll of the dice what platform the Marine aviator would ultimately fly. The top graduate in the class, however, always got his or her first choice. At that point, I had never really done better than average, so graduating at the top of my class was something that seemed impossible. But on my first helicopter flight, I imagined tipping the aircraft over and launching rockets from a Cobra, and I was hooked. In my mind, I had to fly the Cobra. I now had a vision and a goal, and I knew I was going to have to do a lot better than average to achieve it. Instantly, my grades in flight school took a drastic turn for the positive, and I graduated third in my class. During the week that I finished, there was only one Cobra slot available; yet, I was the only Marine

---

[3] The UH-1N helicopter is commonly referred to as a Huey.

in that class who desired to fly the Cobra. So, I ended up with my choice: Cobras on the West Coast at MCB Camp Pendleton, California.

THE COBRA

The AH-1W Cobra attack helicopter is an awesome flying machine. I was assigned to Marine Light Attack Helicopter Squadron 169 (HMLA-169), also known as the "Vipers." The HMLA community is known throughout the Corps for being a very tough community. I think I was developing an ulcer during that first year from trying to adjust to the culture. In the Cobra community, if pilots do not bring their best games, they are quickly discarded. The amount of studying I did in my six years in the Vipers was unparalleled for me.

Flying an aircraft is one thing; using it to fight is altogether another thing. I was struggling my first year in the Vipers when a captain, who saw potential in me, pulled me aside and mentored me. He used an analogy: "'Gonzo' [my call sign], if you were a brain surgeon instead of a Cobra pilot, would you want to perform open-brain surgery on your wife if she needed it, or would you want the best brain surgeon in the country to do the surgery?"[4] As I was thinking about my answer, he asked, "To put it a different way, if you were responsible for cracking someone's skull open and digging around in his brain, would you be studying harder than you are now as a Cobra pilot?" I responded that I probably would be studying a bit harder if I was going to actually perform brain surgery, and then the tone changed immediately. It would be inappropriate to use the colorful language that ensued, so allow me to summarize the lesson: there is no difference between the responsibilities of a brain surgeon and a Cobra pilot in terms of preparation and experience. Lack of knowledge and mastery by a brain surgeon will just as assuredly get a patient killed as a Cobra pilot could get Marines killed without professional dedication to mastering our craft. As Cobra pilots, we should seek to be so skilled, knowledgeable, and

---

[4] Although likely intended early on to allow ground controllers to easily communicate with pilots, call signs today are created by the members of air units for each pilot and usually reference their names, physical traits, or specific events.

proficient in the craft that others seek our expertise. My confidence needed to be equal to the brain surgeon who would demand that he, and only he, be the one to perform surgery on his wife. Lesson learned. And from that moment on, just like in flight school, I had a new motivation. No Marines were going to die due to my lack of preparation. I was going to be the best Cobra pilot that I could be, and no one was going to out work me. My life for the next five years almost entirely consisted of 12-hour days at the squadron and studying until I fell asleep at night.

Then, the unspeakable happened on 11 September 2001, while I was a captain on deployment in Okinawa. I was up late studying weapon systems, with the TV on for background noise. When the second plane hit, I knew we were going to war. When we returned to Camp Pendleton, I attended the Weapons and Tactics Instructor (WTI) Course. Between the awesome instructors I had with the Vipers and the incredible training and instruction I received at WTI, I could not have imagined being better prepared for flying in combat. I owe a lot to the many instructors I had in my career and to the challenging yet realistic training I received. I also owe a lot to my peer group in the Vipers. We had a very close-knit group that had an incredible balance between friendship and intense competition. We pushed each other to be the best we could be. Measuring up to their talent level was another factor that helped to drive me to work as hard as I did. They were, and are, the most competent group of people I have ever known.

When it became clear that the Marines were going to Iraq, and I MEF was getting the assignment, our attention became a kind of myopic panic over which light attack helicopter squadrons were going to go, or more importantly, who was going to get left behind. As it is stated in the 1995 Marine Corps Warfighting Publication 6-11, *Leading Marines*, "Marines are going in harm's way, and there is an unnatural feeling of being 'left out' among those not able to go. . . . One Marine father sending his son into the Corps summed it up this way: 'May our Corps not have to go in harm's way on your watch; but if it does, may you never be the second Marine there.'"[5]

---

[5] Marine Corps, *Leading Marines*, MCWP 6-11 (Washington, DC: Headquarters Marine Corps, 1995), 16.

Yet, the reality was that some Cobra pilots were going, and some were not. HMLA-169 Vipers, and HMLA-267 "Stingers" were the two squadrons from Camp Pendleton that went to Kuwait for OIF I. HMLA-367 "Scarface" was still on deployment in Okinawa. And so, it was HMLA-269 "Gunfighters" that would not go to war. I felt for my friends in the Gunfighters. They were a great squadron, and being left behind while fellow Marines went into harm's way was unthinkable. To be honest, every light-attack helicopter pilot thinks that his or her squadron is the best that ever existed and, thus, should be the one to go to war. That is probably a good and healthy confidence to have. We certainly felt we were, but regardless, I was relieved it was not my squadron that had to stay behind. It seems silly now, because at the time, we had no idea we would rotate squadrons through Iraq and Afghanistan for the next 10 years (and counting).

We traveled to Kuwait via a Lockheed C-5 Galaxy airlift and spent about three weeks practicing tactics with our Cobras and Hueys, drilling for chemical attacks, and building up the expeditionary airfield at Ali al-Salem Air Base with AM-2 matting.[6] It was pretty slim pickings as far as the accommodations went. I recall the "grade F" burgers being served up at the chow tent in boxes labeled "WARNING: For Prisoners and Military Only!" The speaker system we called "Giant Voice" would echo the warnings for practice drills a couple times per day. We were practicing full MOPP gear all the time, expecting a chemical attack via a Scud-B surface-to-surface missile.[7] The fighting started on the evening of 20 March 2003, and on 21 March we had nine actual incoming warnings. U.S. Army Patriot antimissile batteries were fired off to intercept inbound missiles. I do not recall any ever coming very close to hitting our air base, but the enemy forces tried.

The next month and a half was a whirlwind, a blur of memories, but some memories are still pretty solid in my mind. I recall flying to the border

---

[6] Ali al-Salem is a Kuwaiti air force base located west of Kuwait City. See Col Nicholas E. Reynolds, *U.S. Marines in Iraq, 2003: Basrah, Baghdad, and Beyond* (Washington, DC: Marine Corps History Division, 2007).

[7] MOPP refers to mission-oriented protective posture gear or individual protective equipment including suit, boots, gloves, and mask with hood issued to military members in preparation for possible chemical attacks.

of Kuwait and Iraq and looking out to the west and witnessing the Army's 3d Infantry Division (3d Division) staging to cross the line of departure. It was magnificent to behold—horizon to horizon, M1A1 Abrams main battle tanks, Bradley fighting vehicles, trucks, and equipment. I had been flying over the Marines' positions for three weeks and could not recall seeing even one-third the number of vehicles, particularly tanks. It was striking to see the differing approach to warfare so apparently displayed from 300 feet above the ground, the best view in the world. The contrast was clear: strong and steady versus light and fast. On the left, the Army's 3d Division resembled the bullmastiff; while on the right, the 1st Marine Division was more like a Doberman pinscher.

On 21 March, after flying over the al-Faw Peninsula and supporting the British in al-Basrah, we were directed to support the lead Marine elements moving north toward Baghdad. The forward air controller (FAC) directed us to his position: "I'm in the tank with the orange signal panel; call contact."[8] As we flew directly over him at about 50 feet, we called, "Contact." "I am the lead tank," he said. "Anything military north of me is bad guys. If it isn't waving a white flag—kill it." Proceeding north, out in front of everyone, armed to the teeth, into a country that did not want us there was a surreal feeling for me. We spotted groups of Iraqis clearly surrendering and called out their positions to the grunts. The war equipment they were abandoning, we destroyed. My front-seat copilot, who was as green as could be, had never fired an actual AGM-114 Hellfire air-to-surface missile. His first and second Hellfire missiles destroyed a ZSU-23-4 Shilka antiaircraft gun and an SA-9 Gaskin surface-to-air missile system that we found hidden under a bridge.

As I remember it, the battle rhythm (if there was one) went as follows.[9] We departed Ali al-Salem with a division of four Cobras and flew missions in support of the Marines on the ground for about three days.[10] We slept on

---

[8] FAC is an aviator or pilot who is a member of the tactical air control party who, from a forward ground or airborne position, controls aircraft in CAS of ground troops.

[9] As used here, battle rhythm is a deliberate daily cycle of command, staff, and unit activities intended to synchronize current and future operations.

[10] A division consisted of four Cobra attack helicopters.

the ground under our aircraft in the forward arming and refueling points (FARPs) that were established as the infantry moved forward. There is very little storage space in a Cobra, so we stuffed the small opening in the tail boom with compressed sleeping bags, changes of underwear, our chemical masks, and some meals, ready to eat (MREs). Under my seat, I stored as many water bottles as I could. We carried M4 rifles and 9mm pistols in the cockpit, along with our NVGs. Generally, the day started early, at sunrise, and we started the aircraft and contacted "Skychief," the direct air support center (airborne) (DASC[A]).[11] Skychief directed us to wherever we were needed with a grid, a frequency, and the call sign of the FAC to contact. We supported the FAC with whatever he needed, usually in the form of CAS, and repeated the process by cycling in and out of the FARP for fuel and ammo without shutting down the Cobras all day. The most mundane details turned out to be what left a lasting impression on me. The Cobra is a tight fit on the inside and provides a rather uncomfortable ride, which was tolerable for short flights. We were not, however, accustomed to flying them for 8–12 hours straight. Often, we started the aircraft at about 0600 and did not shut down until sunset, only getting out of the cockpit during refueling for about five minutes each time for a chance to stretch and relieve our bladders. I recall doing pseudo wall sits in the cockpit in an effort to get my aching a——s off the seat. I began taking Motrin because my legs and back were in such pain from sitting in that uncomfortable and tight space for so long. I remember thinking to myself at the time how funny it would be if someone in the future asked me what it was like to fly in combat, and I could honestly respond: "It's a real pain in the a——s!"

After a few days of flying missions all day, sleeping on the ground, c——ping in the dirt, and milking the few MREs we had to make them last, it was the aircraft, not the pilots, that needed to be taken back to Ali al-Salem for inspection and maintenance. We would fly them back, get sleep, and, if we were lucky, get a shower (nothing ever worked in Ali al-Salem, includ-

---

[11] DASC is the principal air control agency of the Marine air command and control system responsible for the direction and control of air operations directly supporting the ground combat element.

ing the showers), and then, we would go back out for another three to four days of work.

My division of Cobras was on a second rotation in Iraq when we spent most of one day supporting Marines pushing through the city of an-Nasiriyah.[12] When we showed up on station, we relieved the division that had been there through the night during the infamous ambush on the U.S. Army convoy where Private First Class Jessica D. Lynch was captured.[13] We worked with the Marines on the outskirts of the city to take out some sniper positions across the Euphrates River, and we tried to do some reconnaissance for the grunts over "Ambush Alley." As we flew over the city, I heard the FAC's garbled voice over the radio shouting, "You're taking fire!" The sky suddenly lit up with antiaircraft artillery blasts. Large black puffs of smoke about the size of a small car blasted all around us. We were pretty lucky not to get shot down, because the Iraqis had our distance and elevation dialed in; they just missed us. Recon (reconnaissance) over Ambush Alley on that day was untenable.

THE CONVOY

It was still March 2003 when our division, on its third rotation into Iraq, was directed back to an-Nasiriyah. Our division of four Cobras was given the call sign "Clancy 31." For this story, I will only identify pilots by their call signs. The division leader (dash-one) for Clancy 31 on that day was "Penguin;" his copilot was "Sweaty."[14] Penguin's wingman (dash-two) was "Nuts;" his copilot was "Happy." The second section lead (dash-three) was "Gecko;" his copilot was "Woody." I (Gonzo) was Gecko's wingman (dash-four); my copilot was "Gorby."

---

[12] An-Nasiriyah, a city in southern Iraq, was the first major challenge for the Marine's Task Force Tarawa, which initially needed to secure the city and the bridges over the Euphrates River. See Reynolds, *Basrah, Baghdad, and Beyond.*

[13] Lynch was a member of the Army's 507th Maintenance Company when she was taken prisoner by Iraqi forces on 23 March 2003. She was rescued on 1 April 2003 by U.S. Special Forces. See Reynolds, *Basrah, Baghdad, and Beyond.*

[14] The terms *dash-one, dash-two*, etc., refer to physical position of the aircrafts in formation. Dash-one is used for the lead aircraft.

Before we got orders for a convoy mission, we landed along a small road where a new FARP called Camden Yards had been created just south of an-Nasiriyah. This FARP was really about as basic as it could get: a fuel truck, a couple of ordnance trucks, and a handful of maintenance and ordnance Marines. We had to sleep in a ditch that night next to our Cobras. It was pretty cold for just a sleeping bag liner and what we were wearing. We received our first mission at about 0630 on 28 March from Skychief. We were to provide security for a Cobra that had run out of gas and had autorotated to the desert floor.[15] A CH-46 was inbound to bring it fuel so it could get back into the fight. We located the Cobra and coordinated the recovery, and then, we got the convoy mission.

At about 0730, Skychief came on the radio with orders, "Clancy 31, contact 'Housefly' or 'Chilipepper' and coordinate to provide convoy escort from Camden Yards to Regimental Combat Team 1 (RCT 1)." Housefly was the call sign of the convoy commander, and Chilipepper was the FAC with a light armored reconnaissance (LAR) unit, providing ground escort. After refueling the Cobras at Camden Yards, we located the convoy not far from our position. You could not miss it (it was a big convoy), and it was heading for an-Nasiriyah. Penguin, our division leader, landed his Cobra in the dirt next to the ground escort commander. Penguin got out of the Cobra and coordinated face-to-face with either Housefly or Chilipepper or both. They had to figure out a plan on how we were going to tackle this, because this was no ordinary convoy of 30–40 vehicles. This was going to be a convoy with more than 200 vehicles!

The infantry was moving north fast and needed the ability to establish more FARPs so rotary-wing support could be close at hand. The original convoy was supposed to only consist of FARP and rapid runway repair elements of Marine Wing Support Squadron 371 (MWSS-371) and 12 LAV-25s from LAR for ground escort. This totaled about 100 vehicles, but due to the fighting in an-Nasiriyah, the vehicles were delayed by two days. After Task Force Tarawa had secured the bridges and Ambush Alley, the Marines

---

[15] Autorotated refers to a controlled landing maneuver when a helicopter experiences engine failure.

were ready to move north when the force service support group sent an additional 100 vehicles to piggyback onto the already enormous convoy.[16] It is not surprising that a 215-vehicle convoy would get the attention of someone directing aviation assets, and that is how we came to pick up the mission of escorting the convoy from Camden Yards north to RCT 1.

Penguin ran to each of our aircraft and pointed out on the map where we were going and what each of our responsibilities would be within the division. We were going to escort the convoy all the way north to RCT 1 so they could set up Fenway FARP by Qalat Sikar. We pushed ahead of the convoy and made our presence known over an-Nasiriyah, flying over the exact same spot we were nearly shot down at just days before. It was weird seeing an-Nasiriyah as calm as if no one was there, when just days earlier, it was a hotbed of enemy fighting. We got the convoy through to the north of an-Nasiriyah without anyone taking enemy fire. It takes a long time to get 215 vehicles moving in the same direction, so by the time they got through an-Nasiriyah, we were bingo on fuel (out of gas) and had to head back to Camden Yards. Once we were refueled, we were temporarily retasked to detach from the convoy and perform armed reconnaissance in support of Task Force Tarawa.

Sometime around 1200, we received an immediate air request from the DASC, whose call sign was "Chieftain," to proceed north along Route Bismarck (Highway 7) and contact the convoy. We contacted Chilipepper, who reported that the convoy was ambushed in the vicinity of a small town called ash-Shatrah with small arms and rocket-propelled grenades (RPGs).[17]

---

[16] The 15th MEU (Special Operations Capable) [SOC] provided a security force to run the 200-vehicle convoy through an-Nasiriyah to Qalat Sikar. While en route, the LAR company escorting the convoy encountered heavy resistance and defeated deliberate ambushes along Iraq's Highway 7 and through a four-kilometer stretch of road connecting the bridges over the Euphrates River and the Saddam Canal known as Ambush Alley. LtGen Wallace C. Gregson, "I Marine Expeditionary Force Summary of Action," in *U.S. Marines in Iraq, 2003: Anthology and Annotated Bibliography*, comp. Maj Christopher M. Kennedy et al. (Washington, DC: Marine Corps History Division, 2006).

[17] See Commander and Staff, Marine Wing Support Group 37, "No FARP Too Far!," in *Anthology and Annotated Bibliography*; Nathaniel Fick, *One Bullet Away: The Making of a Marine Officer* (Boston: Houghton Mifflin, 2005); and Jay A. Stout, *Hammer from Above: Marine Air Combat Over Iraq* (New York: Presidio Press, 2005).

Chilipepper described the ambush site as a building next to the highway with arches and several pictures of then-Iraqi President Saddam Hussein on the west-facing wall and gave us type-three clearance to find and engage the suspected ambush site.[18] We approached from the west, as we did not want to just barrel up the obvious avenue of approach. Ash-Shatrah was a small town, mostly built up on the east side of Highway 7. There was a brick factory just to the west, easily distinguishable by the two large smoke stacks (possible obstacles to low-flying helicopters) and some wires to be concerned with if we got really low. We utilized the optics in the Cobra to locate the building described by Chilipepper, which was suspected to be the local Baath Party headquarters, and sure enough about a dozen enemy fighters were there carrying small arms, so we sent them a couple blast-frag Hellfire missiles and destroyed half the building.

We returned to the convoy, passed the battle damage assessment (BDA) to Chilipepper, and then had to return to Camden Yards for more fuel and to reload our missiles. Once complete, we flew back north to relocate the convoy, and to our surprise, the whole convoy had turned around and was headed south. After what seemed like an eternity of flying butterfly patterns—one section on the east side of the highway, one on the west side, 180 degrees out from each other—the convoy turned back north and proceeded to RCT 1 at Qalat Sikar as originally planned. This made us very happy, because we did not want to try to get those vehicles through an-Nasiriyah at night, which was quickly descending upon us. Well, it took a long time to coordinate 215 vehicles to turn around and get moving, and as we approached ash-Shatrah, we were running out of fuel again. When we told Chilipepper we needed to go back to Camden Yards for fuel, I could tell in his voice that he was not happy. I think his exact words were, "What? You can't leave us!" Penguin, thinking out of the box, said, "I see a number of fuel trucks down there, we could land on the road and have them refuel us right here." This was totally unorthodox. To do this meant refueling armed

---

[18] According to the 2014 Joint Publication 3-09.3 on CAS, type-three clearance is the least restrictive for air support and gives the aircraft authorization to engage enemy on the ground, who may or may not be engaging friendly forces.

Cobras while they were running on the side of the road without being in a sanitized FARP. Chilipepper said, "Can we do that?" and quickly followed with "Stand-by Clancy. I'm coordinating." He and Penguin continued to work out the possibilities. I cannot recall all of the coordinating that took place, but at one point Chilipepper, arguing with someone, said, "If we don't refuel them here and now, they will have to leave us and go back to Camden Yards for gas." There was a long silence and then a loud and direct, "We're making a hole for them right now!" Clearly, the convoy did not want to be left alone.

The vehicles left a gap in the convoy, just enough room for two Cobras, so we came down two at a time and got the gas we needed for the next two hours. I was in the second round of refueling and had Gorby get out to disarm the rockets and supervise the Marine putting fuel in the aircraft. It was quickly getting dark, so I took this time to get my NVGs on and ready. The fuel truck pulled up, and Gorby gave me a thumbs-up that the refueling was going well. I looked back at the young Marine refueling the aircraft, who was close to me. The gravity fuel port is right behind the backseat pilot on the right side, so he was almost within arm's distance from me. As I looked back, he looked up at me with wide eyes. I could not read his mind, but I thought he looked scared, and at that moment, I could not blame him. Marines in the convoy had already been ambushed; they were about to go through the same ambush site, only this time in the black of night; and he was driving a giant fuel truck!

That was a moment of clarity for me about the importance of my duty as a Marine aviator. That one Marine (maybe 19 years old) was the representation of whom I was training my whole career to protect. He was young, he was driving a big target, and he needed us. I had looked that Marine in the eyes and saw the reason I was there. He might have needed the best "brain surgeons" possible during this operation. I gave him a reassuring nod and wink. He looked at me and recognized I was communicating with him, which I think took him by surprise, and he slowly nodded back. The nonverbal, Marine-to-Marine expression of "I've got your back" was communicated and acknowledged. His wide-eyed look hardened, and he turned his focus back to filling up the turning Cobra.

We got back in the air and, within minutes, were in full-blown, low-light level conditions. To nonaviators, that translates to flying in the dark under difficult conditions. While we were refueling, Penguin and Chilipeper had coordinated a plan for us to push ahead of the leading LAVs arriving in ash-Shatrah and hit those buildings at the ambush site again. We set up from the west and launched a couple more Hellfire missiles into the buildings as planned preceding the LAVs. Once complete, we set out to reestablish the butterfly pattern over the convoy for optimal protection in case the enemy was bold enough to attempt another ambush despite knowing four Cobras were overhead.

At that time, another division of Cobras from HMLA-267 showed up. Apparently, the DASC was pushing air assets our way to help escort the convoy, and that turned out to be a good thing. We coordinated that the division from HMLA-267 would cover the back half of the convoy while we, the division from HMLA-169, covered the front half, which was more than enough work as this convoy stretched out for miles along Route Bismarck. The 12 LAVs providing ground escort were divided into three groups: four LAV-25s in the lead, four in the center, and four at the tail end. As the lead vehicles got to the ambush site, they immediately came under heavy fire from multiple positions within the town of ash-Shatrah. All the buildings were on the east side of the road, and heavy fire was coming from about a block's worth of buildings firing east-to-west into the LAVs. The glowing tracer fire came from the buildings, hit into and around the LAVs, and ricocheted up into the air in all directions. Almost instantly, the LAVs opened fire on the enemy positions, and that blazing tracer fire went west-to-east into the buildings and ricocheted up intermixing with the enemy fire filling the sky like a pyrotechnic firework display. Penguin communicated with Chilipeper, trying to get clearance to attack the enemy positions. Chilipeper was positioned with the LAVs in the middle of the convoy about a hundred vehicles back from the front of the convoy. The trust he must have had to clear us to attack the enemy was enormous. Penguin painted the picture for Chilipepper quickly, letting him know this was a linear target array with easily identified enemy and friendly positions. This was going to be danger close, so he was clearly nervous about clearing us, the Cobra pilots.[19] He wanted reassurance that we knew exactly what we were going to be shooting at. Penguin picked up on this and confidently and clearly stated, "Clancy is tally enemy, visual friendly.[20] We will maintain our attack along the north-south road axis." Chilipepper understood and responded, "Clancy, you are cleared type-three control to attack enemy positions." With that, Penguin came up on the interflight radio and coordinated our attack: "Trail attack, left-hand pulls." He did not have to explain that Marines were within 100–200 meters of the enemy and that we would be pushing in closer than we had ever pushed in before. One errant rocket and we would be doing the enemies' job for them. We had to get close to be accurate, and we had to fly into that heavy fire to do it. The division simply responded: "Two." "Three." "Four." We all got it.

Penguin dove in. I could tell he was shooting because flashes from his rockets were causing my NVGs to wash out. NVGs are light intensifiers and are very sensitive. When confronted with bright light, the goggles will go blank from the intense light, and it takes a second for them to recover. With all the tracer fire, the rockets, the city lights, and the explosions, it became very disorienting in the low-light level conditions. In the dash-four position, I focused on not running into Gecko in dash-three, as well as maintaining contact with the enemy and friendly positions. I knew Nuts, in the dash-two position, was now firing, because of the blinding light from his rocket fire and impacts. Gecko made his attack run, and I set up to have my nose on the target ready to fire the second he cleared the target area. I wondered for a moment if there would still be enemy left after three Cobra attacks, but it was clear that there was as the rate of fire from both sides was still very heavy. With tracer ricochets flying up and in my face, I picked out the source of fire from one of the enemy positions and pushed in, I think

---

[19] *Danger close* is a term included in the method of engagement segment of a call for fire that indicates that friendly forces are within close proximity of the target.

[20] *Tally* means that the enemy position or target has been positively identified.

I shot the first rocket at about 300 meters. It was on target, so I followed it up with two more. That was all I had time to shoot. Being that close, I had to pull off or run into the target. "Back stick, power, turn." was the mantra drilled into my core from those who trained me. Now, up and away from the target, I needed to quickly locate Gecko. Predictability is key to avoiding a midair collision in these conditions, and Gecko was right where I thought he would be, but he was hard to see because of all the tracers flying past my aircraft. Between Gorby and myself, we maintained contact with Gecko and heard Penguin call out another attack. As we were coming around for a second pass, I saw the flashes from Penguin's rockets in my peripheral vision, and then the volume of tracers flying past my aircraft lightened up. I stayed in position in trail of Gecko, and everything happened as expected, just as we had done countless times in training. We kept track of each other using minimal communication: "Lead is off left, visual four." Then: "Two is in," which was followed by flashes of light from his rockets. "Two's off left, visual one." And "Three's in." Gecko's Cobra tipped into the ricocheting fire, and his rockets pounded the targets: "Three's off left, visual two." As the fourth Cobra, I now tipped in again, looking for the source of enemy machine-gun fire. I was happy to see only one gun still firing on the Marines. I lined up on that source, got a visual of the Marines, eyes back to the enemy, and made my call: "Four's in." I fired four rockets into that machine gun. Back stick, power, turn—find Gecko, got him. "Four's off left, visual three." This time, there was no tracer fire blasting past my aircraft as I pulled off target. I looked back and all the enemy guns were silent.

We stayed with the convoy as it traveled north along Route Bismarck. The next small town along the road, about 20 kilometers north, was an-Nasr and was eerily similar in layout to ash-Shatrah. As the lead elements of the convoy reached an-Nasr, the back half of the convoy was still passing through ash-Shatrah. When the lead LAV reached an-Nasr, it immediately came under fire. It was exactly the same as ash-Shatrah. The radio chatter picked up, and Penguin was attempting to coordinate with Chilipepper, but it got very confusing because, at the same time, the rear of the convoy still passing through ash-Shatrah came under fire again. Both divisions of

Cobras began CAS attacks on the enemy positions, much the same as we had just done. As I pulled off target from my first pass on enemy positions and began my left-hand turn, keeping interval off of Gecko and Nuts, two large rocket-like projectiles launched from the ground and blasted right through the middle of the division. That got our attention, but we decided to stay focused on the immediate threat to the convoy and made a second pass from north to south on the enemy positions firing on the Marines. We proceeded as we had in ash-Shatrah; two passes were enough to silence the enemy machine guns and RPGs. With a break in the fire, we engaged the launch site just for good measure. The back of the convoy was experiencing some chaos, and from the radio chatter, we could tell that multiple vehicles were hit and some had gone off the road. The Marines in the convoy were aggressively attempting to get accountability as they continued north.

We were once again out of gas and out of rockets. As we began our return to Camden Yards (now much farther away), two Marine F/A-18s were checking in to continue to escort the convoy while we were gone. We were asked to use our sensors to do recon along the road at the rear of the convoy as we headed south back to Camden Yards. We could not spend much time searching due to our fuel state, but we did see a vehicle, as best we could determine it was abandoned, and we could not find any Marines near it—dead or alive.

Once again, we got fuel and ordnance from Camden Yards and headed north back up Route Bismarck, past an-Nasiriyah, past ash-Shatrah, past an-Nasr, beyond a very small town on the map called ar-Rifa'I, and on toward Qalat Sikar where the last elements of the convoy were nestled in with RCT 1. We contacted Chilipepper, and I recall him saying, "We had a feeling you'd be back to make sure we made it." They had made it, and we were cleared to go back for the night. We had started that morning at 0600 and shut the Cobras down in Camden Yards at about 0200 in the morning the next day—20 hours later. We looked over our aircraft as best as we could in a moonless, dark night, but planned to look them over for bullet holes again more closely in the morning light. We conducted a short debrief and then slept in the ditch along the small road under the tail boom of the Cobras.

I remember having a hard time walking because my a——s and back hurt so much. Laying down on the ground did not help, but it was nice to be on terra firma. The next morning we were all very surprised to find out that not one Cobra had battle damage, so we fired up the aircraft and contacted Skychief: "Four Cobras, up as fragged."[21] Another day's work ahead, and so it went.

THE CRISIS

Word of the convoy escort got around. We lost one Marine. Literally "lost" him. We were asked to look for that Marine along the route our last time back to Camden. Our fellow squadron mates heard about the mission when we finally got back to Ali al-Salem to get a couple day's rest and some new aircraft. We were all sharing our stories and trying to learn from our collective experiences. The 1st Marine Division (1st MarDiv) had paused from the rapid drive north to clear their rear areas and prepare, in coordination with the Army, for the final push into Baghdad. It was at this time, the grunts went back through ash-Shatrah to look for the lost Marine, and they were able to get a BDA from the fight along Highway 7.

It was dark, and I was coming back from a tent where we had a TV showing CNN when a couple of guys approached me. "Gonzo, we just got word that they recovered the lost Marine's body from that convoy mission."[22] This was not good news. For despite knowing he had probably been killed, we were all hoping they would find him alive. "They also said you guys killed a lot of *fedayeen* fighters." With a sense of dread, I asked, "What do you mean by a lot?" One Marine informed me that, "They don't have a number. They said the Iraqis had already buried them, but that there were a lot of them." And that is when the sinking feeling came. I can tell the

---

[21] *As fragged* means the fighter, FAC, mission package, or agency will be performing exactly as stated by the air tasking order.

[22] Sgt Fernando Padilla-Ramirez, assigned to MWSS-371, was reported missing in action on 28 March 2003 as his convoy traveled near an-Nasiriyah. The 24th MEU (SOC) recovered his remains on 10 April 2003. LtCol Melissa D. Mihocko, *U.S. Marines in Iraq, 2003: Combat Service Support During Operation Iraqi Freedom* (Washington, DC: Marine Corps History Division, 2011).

next part of this story a hundred different ways, and none would be able to convey what I was feeling. This story is not only about what it is like to be in combat, but it is also about what someone goes through after combat. Some people simply will not understand it. To be honest, I would not have understood it had another Marine approached me at that time and expressed feeling similarly. But having lived through it myself, whether rational or irrational, this is my best attempt to describe what I went through.

I was suddenly having an ethical crisis. Getting unequivocal evidence that I had taken human life was the worst day in my life. Period. I have an impression in my memory, almost in slow motion, of squadron buddies high-fiving me; my response was automatic because my mind was not celebrating. I was in a sudden fog. The scene often portrayed in movies that depicts a person experiencing tunnel vision with all the eerie music is the best way I can describe what I was going through. Years of learning Catholic theology rushed to my consciousness. "Thou shall not kill" boomed in my mind overlapping a foggy impression of giving high-fives. A distorted voice in the background added to the experience saying, "Way to go. You killed a bunch of bad guys!" An image of Father Lutz flashed into my mind. He was crying.

I had to get away. I could not let anyone see me questioning what I had done. I quickly made my way to the chaplain's tent where they were just beginning an evening Mass. "Thank God," I thought and went right in. I did a lot of praying. I did not take Communion, because I did not know if that was the right thing to do, at that point, with *that* on my hands. I repeated one prayer over and over: "God, if what I'm doing is wrong, somehow let me know. I need you to talk to me." I needed a definitive moment of clarity, but all I could think about was that same feeling I had when talking to Father Lutz. Jesus would not be doing this. I told myself I had to get my head right because I was going back in the saddle in two days. Marines would be counting on me, and I had to have my head in the game. I kept thinking: why was I not prepared for this? I can say, from experience, that one can intellectually rationalize every combat scenario and be very comfortable with what might happen in combat—the potential for experiencing a grizzly death or the reality that some actions must be done that cause harm to others. Here I was, multiple combat missions under my belt, and suddenly, with the news that I had killed "a lot of bad guys," emotionally everything had changed. What I am saying is that rational analysis and years of conditioning cannot always predict the emotional response to information, information that was not even new or unexpected. I did not know I would have such an emotional response when confronted with that. I thought I had intellectually come to grips with the possibility of the reality of war. It would not have been right or fair to the Marines in my squadron, or the Marines we were protecting, to continue if I had not.

Most of us are both rational and emotional beings. Obviously, the former can be dissected, explained, and quantified. The latter . . . well, who knows? What I know is that most of you reading this are probably thinking what I was thinking as well: you are a trained Marine Cobra pilot. You had to expect this, right? You had to be prepared for this, right? You knew what you were doing. High-explosive rockets are designed to kill, and you launched them, accurately, into enemy positions. That was your job, you did it, and you saved Marines lives. There was no ethical dilemma. There was no gray area involved here. Mistakes were not made; innocent people did not die. It was a black-and-white scenario—kill or Marines get killed. This was all rationally true. Yet, my brain was taking me out of the tactical situation, and I was suddenly questioning eternity and salvation juxtaposed with my actions. After Mass, it was dark and I was walking back to my tent. I was still struggling with reconciling my faith and my profession. I thought again: "God, I need you to talk to me." Just then I came across one of the pilots in my division.

We did not have enough pilots at the time of the deployment to fill all the seats in the aircraft, so we needed at least five pilots from other squadrons to meet the missions. Sweaty was one of those pilots. He was flying as Penguin's copilot. Sweaty was well known as a Bible-thumper. He was one of the nicest guys I had ever met. He did not swear or drink, so as you might imagine, he was a bit of an odd duck in the Cobra community. Sweaty was just sitting there in the starlight relaxing outside his tent when he saw me and said, "Hey Gonzo, everything okay?" I stopped and stared at him. He should not have even been in my squadron let alone in my division. But here he was, sitting right there as I asked God to talk to me. "Bible-thumper," I thought as I looked up at the stars and back to Sweaty. No way. I must have stared an inquisitive hole into Sweaty's forehead. I do not know what I was looking for? A halo? Wings to sprout from his back? I shook my head like I was shaking out of a hallucination and began to walk past him. "Gonzo! You okay?" I stopped and thought, "Well that's twice. Are you going to walk away from what could literally be God answering your prayers?" I turned and stared at him, wondering if I should risk talking to him about what was bothering me. "Okay, God, if this is your way, I'm going to trust you," I thought to myself before sitting down next to Sweaty. I said, "You're a Bible guy." Sweaty laughed, "Yeah, I'm a Bible guy." With that, I asked the big question, "How do you know what we are doing isn't against what God wants from us?" It became apparent that Sweaty was there for a reason. Sweaty started quoting scripture, most of what he said I recalled reading myself or hearing at Mass. None of it was new, but talking with Sweaty helped me access memories, stories, and lessons I had forgotten. It was just what I needed. We talked for about 45 minutes, and when we were finished, I was good. Weight lifted. Prayers answered.

Emotionally and physically exhausted, I lay down on my cot to crash for the night. I thought about everything that had transpired: the convoy, my moment of doubt, my prayers for answers, my chance run-in with Sweaty, and our long conversation. I said a prayer for everyone fighting, both the enemy and friendly forces. That no matter whose numbers were getting punched, they would be right with the Man. I thought about Father Lutz and the last time I saw him. I thought about the image of him that popped into my head in my moment of doubt. Then, I realized that image was not some new image of him crying because he was disappointed. It was the same image of him blessing me and crying, and now, I understood. He was crying then because he knew I would struggle with this, and when I was struggling, his face popped into my mind. To this day, I wonder if it was an amazing example of how the subconscious mind works or if it was just Father Lutz trying to talk to me in a way that is beyond words.

It was months later when Marines from the convoy, retrograding back home, made their way through Ali al-Salem. I was not there. I was in Iraq on strip alert standing watch over the last Marine units there.[23] Some of the Marines from the convoy went out of their way to find the Cobra pilots that escorted them that night, but we were scattered all over the place flying missions. However, they did find Gecko. I am told they all shook his hand, thanked him, and told him that without the Cobra support they would have had significant losses. I admit that I have a degree of peace knowing that they said we saved a lot of Marines that night.

This reminiscence is dedicated to Sergeant Fernando Padilla-Ramirez from MWSS-371—the Marine who was killed the night that my division watched over the convoy in our Cobras. He was a Marine I never met, but wish I could have.

REFLECTION POINTS

• How is the Marine Corps concept of brotherhood useful in battle?

• How do air personnel adapt to maintaining cohesiveness when they are confined by aircraft and separated from troops on the ground?

• In the case of Lieutenant Colonel Beyer, how did having spiritual faith both help him and distress him? Have servicemembers always had to balance fealty to their nation and brothers-in-arms with their faith?

• How do the results of battle differ for those who experience events first-hand and those who make the orders? From those in the field to those back in Washington, DC?

---

[23] Strip alert is when aircraft are placed in a standby or on-call status.

CHAPTER FIVE

THE BASTARDS' SHEPHERD:
A NAVY CHAPLAIN WITH THE 2D BATTALION,
4TH MARINES, IN AR-RAMADI, IRAQ
COMMANDER BRIAN D. WEIGELT, CHC, U.S. NAVY

In November 2003, Brian D. Weigelt left Coquille, Oregon, to begin his career as an active-duty U.S. Navy chaplain. He reported for duty at 1st Marine Division in January 2004 and, six weeks later, deployed to Iraq with the 2d Battalion, 4th Marines. Over the course of seven months (February–September 2004), Lieutenant Weigelt supported more than 1,000 Marines as they patrolled the streets of ar-Ramadi, the capital of al-Anbar Province in Iraq. His transformation from pastor of a civilian congregation to chaplain of Marines in combat occurred at lightning speed as he shuttled between three bases in ar-Ramadi. He ministered to one of the hardest hit Marine battalions in the early years of OIF. Since leaving 1st Marine Division, Chaplain Weigelt has served onboard ships, at the U.S. Naval Academy, in Djibouti, Africa, and for the Joint Chiefs of Staff.

GETTING READY FOR WAR

The alarm went off at 0300. Setting the alarm had not been necessary; I hardly slept. My ride to Marine Corps Base Camp Pendleton, California, would arrive in 20 minutes. I crawled out of the motel room bed, got ready, and quietly walked to the portable playpen where our 13-month-old daughter Eliza lay blissfully asleep. Bending over, I kissed her gently on the cheek as tears began forming in my eyes. I then kissed my wife, Rosslyn, silently through our mingled tears, and I walked out to the waiting car in the parking lot.

Clari helped me load my seabag and pack it into his car. My wife and I met Clari and Sue at the church we had started to attend. Clari was a retired minister, and Clari and Sue's son, James, had been a seminary classmate of mine. We knew it would be too hard for Rosslyn and Eliza to get up so early and drive me, so Clari graciously volunteered to provide transportation. Driving from Temecula, California, to Camp Pendleton the morning of 19 February 2004, we made small talk, but my mind was reviewing the cascade of events that had taken place during the previous eight months.

In early July 2003, Rosslyn and I were the pastors of a church in the small town of Coquille on the coast of southern Oregon. I was a drilling U.S Navy Reserve chaplain, our church was flourishing, and Eliza was six months old. We had been unsettled for the previous few months, uncertain if we should stay in Coquille or move on to another town or have me go on to active duty. Driving home from a trip to Salem, Oregon, we turned to each other and said, "Active duty?" The following week, I left for two weeks of reserve duty, and we determined to pray about whether I should become an active-duty Navy chaplain. It did not make a lot of logical sense—the United States had just gone to war in Iraq, we had a six-month-old daughter, and we had a growing church that loved us. Yet after those two weeks of separation, we both sensed that God was calling us to make this drastic change. On 23 November 2003, we had our last Sunday in Coquille. After four years and four months, we left the community of faith and its members who had taken a risk on young seminary graduates and allowed us to journey through difficult days with them. A series of tragic events caused one saint in the church to say, "Pastor,

I don't know what God is preparing you for, but you've been through more in these few years than most pastors experience in a career."

In February 2004, I was in a car riding through northern San Diego County on my way to Camp Pendleton to deploy with 2d Battalion, 4th Marines, to ar-Ramadi, Iraq. I had checked into the battalion on 5 January 2004 and knew that I would be leaving for a combat deployment in about six weeks. Though I had excelled academically in college and the seminary and though I had been through the Navy's basic chaplaincy and expeditionary skills training courses and though I had developed ministry skills through practical experiences in various settings, I had a strong sense that I was thoroughly unprepared to be a chaplain for a battalion of infantry Marines going to combat. Waves of questions came over me. What programs did I need to implement prior to the battalion's departure? What training did I need to offer the Marines upon our arrival? How would I relate to Marines in such a difficult setting? What had I done by volunteering for active duty?

On my first day at Camp Pendleton in early January, I met the 1st Marine Division chaplain, Navy Commander Bill D. Devine, a Roman Catholic priest. I sat in his office and immediately began expressing my lack of confidence— not exactly a good way to introduce myself to my new supervisor. Exuding incredible pastoral concern, Father Devine graciously said, "Don't worry about programs. Stay close to Jesus—he'll show you what to do." He assured me that if I maintained my spiritual disciplines, immersed myself in prayer, loved the Marines and sailors in the battalion, and listened closely for the voice of God, I would do all the necessary things. In fact, arriving with pre-conceived notions of what must be accomplished may have actually become a hindrance to effective ministry. He also let me know that he would be with me in Iraq, coming by my battalion from time to time to celebrate Mass with our Catholic men and to see how I was doing. His counsel resulted in a great sense of relief for me.

The day of my deployment, we got to the San Luis Rey Gate at Camp Pendleton much too quickly for my satisfaction. We wove from the south-east corner of the sprawling Marine base to the northwest corner, the San Mateo area of Camp Pendleton—home of the 5th Marines, the regiment that famously fought at Belleau Wood during World War I.[1] The 2d Battalion, 4th Marines, was a part of the 5th Marines due to the dissolution of the 4th Marine Regiment. The battalions of the 4th Marines were reassigned to other regiments, and the 2d Battalion then belonged to the 5th Marines. Acknowledging the fact that Marines were in a 2d Battalion to a regiment that no longer existed, 2d Battalion, 4th Marines', nickname is the "Magnificent Bastards."

I first really got to know the Magnificent Bastards during a preparatory field exercise at the end of January 2004. The unit rode in buses from Camp Pendleton to March Air Reserve Base in Riverside County, California. An abandoned military housing complex there was converted into a simulated Iraqi village, complete with forward operating bases (FOBs), role players dressed as Iraqis, and subject matter experts on security and stabilization operations who served as trainers. During the week we were there, I made my home at the battalion aid station with the Navy corpsmen who also served with the 2d Battalion, 4th Marines. Navy Religious Program Specialist Second Class Robert Plaisted could not join me for the field exercise because he was still attached to his previous battalion, with which he had participated in the invasion of Iraq during 2003. It gave me a lot of comfort to know that I would have a combat-tested chaplain's assistant with me in Iraq, but I was disappointed that he would not be joining me for my first real immersion experience in the battalion. Plaisted adopted me into the corpsmen and promised that they would take good care of me in the field.

During those pleasant winter days in Southern California, I got to know men who would become incredibly significant to me in the following year: Lieutenant Colonel Paul J. Kennedy, battalion commander; Major Michael P. Wylie, battalion executive officer (XO); Sergeant Major James E. Booker, battalion sergeant major; Major John D. Harrill III, battalion operations officer; First Sergeant Joseph J. Ellis, headquarters and support first

---

[1] Belleau Wood is a legendary battle for Marines as it was a turning point in World War I when German troops were first halted and then forced to retreat at this site in France. Maj Edwin N. McClellan, *The Unites States Marine Corps in the World War* (Quantico, VA: Marine Corps History Division, 2014).

sergeant; and Navy Lieutenants Kenneth Y. Son and Collin Crickard, battalion medical officers. I had begun the journey of not only being assigned to the 2d Battalion, 4th Marines, but of actually "becoming" their chaplain. It is possible to be assigned as a chaplain to a battalion without ever becoming the chaplain. Becoming the chaplain means that the assigned officer did not squander the goodwill that the Marines readily offered him because of his position. Becoming the chaplain means he has proven that he can be trusted with the deepest pain, most troubling dilemmas, and joyful diversions of his Marines. I began that journey at March Air Reserve Base. By the time I came home from Iraq, I had earned the call sign, "Bastard Shepherd," a name I treasure.

On the day of my deployment, I arrived at the staging area in Pendleton's San Mateo area. Semitruck trailers were parked, ready to receive our seabags and packs. Talking in hushed tones and experiencing the weight of the day, families were gathered in the predawn darkness. I knew, at that moment, that my family had made the right decision about Rosslyn and Eliza staying at the hotel. Clari popped the trunk, and we took out my gear, adding it to the pile. And while we had only met weeks before, he embraced me as though I was his own son. Through his tears, he prayed for me, for my girls, and for my battalion. He promised that he and Sue would take good care of my girls for me and that they would not forget to pray for my protection.

He drove off, and I watched the taillights as long as I could. And then, I turned to the Marines and the families all experiencing their own emotional farewells. It was a mixture of excited anticipation for the Marines—as they were embarking on a journey that they had prepared to go on for a very long time—and the uncertainty of what the future would hold for all of them. It was time for me to move beyond my own feelings of uncertainty, anxiety, and sadness and to embrace the role that I had been called to perform—the role that God had been preparing me to assume. While I will not deny my own emotions, I could not let them control me. It was time for me to be a sign of hope for others. It was time for me to live out what I believed, that we can face any overwhelming and uncertain circumstance with the confidence that comes from knowing God's love and from trusting those around us.

BEING AT WAR

The unit's chartered planes left March Air Reserve Base late in the afternoon on Thursday, 19 February. After brief stops in Maine, Ireland, and Germany, the unit arrived in Kuwait City on Saturday morning, 21 February. We collected our bags and boarded buses that would take us from the modern city to military staging bases in the desert. The flat, sandy terrain extended as far as the eye could see as we drove to the entry control point (ECP) at Camp Victory.[2] This massive complex of tents was filled with U.S. Army soldiers and Marines awaiting the arrival of vehicles at the port of Kuwait, so they could make the journey into Iraq.

The first two weeks in Kuwait were filled with logistics briefs, weapons training, convoy preparations, and physical training. We slept on cots in large tents with plywood floors. It was our opportunity to get acclimated to the desert climate and the constant presence of sand. First Sergeant Ellis had the cot right next to the front "door." Multiple times a day, he would plug his earbuds into an MP3 player, sprinkle water on the floor, and sweep up the sand that had been either blown in or tracked in. I was introduced to the Marine's style of leadership that included taking care of the mundane basics so that the important things can happen without hindrance.

By early March, most of the battalion was ready to travel to Iraq. We had enough HMMWVs and 7-ton trucks to transport the majority of four companies across the desert border. A portion of Echo Company and Headquarters and Support (H&S) Company stayed in Kuwait for another week and then flew into Iraq. Plaisted and I stayed in Kuwait. Father Devine arrived at Camp Victory as the convoy was staging in the desert for a nighttime departure. He was going to hold Mass for the Catholics and suggested that I consider having a Protestant Communion service as well. We rode out into the desert to meet the convoy.

---

[2] Camp Victory, a U.S. military installation in Kuwait, closed in 2006.

After we arrived, we consulted with the leadership and determined when and where our respective services would be. Father Devine asked me to spread the word. Feeling empowered by all I had learned in my months of experience, I went to the companies' first sergeants and told them when and where the services were going to be held. I asked them to spread the word among the Marines, quite proud of myself for knowing how to most efficiently pass the word. I then returned to Father Devine's position and let him know what I had done, awaiting his congratulatory remark on my wise decision. Instead, he said, "Why are you going to rely on the first sergeants to promote your event? You need to tell the Marines yourself." While leading a worship service is important, it is not an end in itself. Relationships are essential, and a chaplain should take every opportunity available to build those relationships—another lesson learned from Father Devine.

After the majority of the battalion began the trek through the desert to the Iraqi border, the rest of us found ways to pass the time awaiting our flight. There were more small-unit level training, more physical training, more relationship building, and more time to consider what we were about to experience. I was going through a roller coaster of emotions, not unlike the others around me. Often, I would wake up with a sense of spiritual emptiness, a sense of loss and uncertainty that was hard to fully describe. At those moments, I had a decision to make. Would I feel sorry for myself, becoming increasingly self-absorbed and isolated, or would I own my emptiness and then rely on my relationship with God to be my source of strength, propelling me into the pain-filled lives of others? It would be a daily decision I needed to face throughout the deployment. And most of the time, I chose wisely.

On Monday, 8 March, with most of the battalion on the way to Iraq, those of us who remained in Kuwait received word that there had been a suicide at Camp Victory. We soon discovered that the victim was one of our Marines. The finely tuned military organization sprang to action with Marines responding to the crisis. Having been the pastor of a small church who had experienced a number of tragedies, my mind raced through a list of all the things that I would need to do. I quickly realized that now I was the chaplain, and all I needed to do was be present to those whose hearts and minds were filled with questions and mixed emotions. Working closely with the battalion executive officer, Major Wylie (the battalion commanding officer was on his way to Iraq with the convoy), my religious program specialist, and I developed a memorial service, which was held the following day, 9 March. The Marines needed to have some way to acknowledge this event and move on—ignoring it because it was a suicide would have only resulted in more uncertainty. At the conclusion of the brief service, the platoon commander of the deceased Marine came to me. The tent was empty except for the two of us. This tall, incredibly fit, young second lieutenant poured out his heart. And for the first time, but certainly not the last, a Marine put his head on my shoulder and sobbed. The honest emotions of Marines—both their joys and their sorrows—captured my heart as I sought to serve them in God's name.

We left Camp Victory on Thursday, 11 March, taking buses to the Kuwaiti military air base near our position. Late in the afternoon, we climbed aboard the airplanes and began our journey to Iraq. We arrived in the dark at al-Taqaddum Air Base (known as TQ).[3] After being told that we would collect the majority of our gear in the morning, we were transported by truck to a large metal warehouse filled with cots. It can be very cold in Iraq in March, and we did not have our gear with us. Needless to say, it was a restless night as we shivered through the night hours. I clearly remember First Sergeant Ellis finding trash bags that he then turned into a field expedient blanket, once again showing me the kind of determination a Marine Corps leader must have in taking control of the circumstances.

I began my mornings—following the advice of Father Devine—spending time in prayer and staying close to Jesus. My prayers were shaped by reading scripture and the book *My Utmost for His Highest* (1935) by Oswald Chambers. Chambers was a Scottish evangelist and teacher born in 1874. Known for a deep personal spirituality and an extraordinary ability to communicate

---

[3] Al-Taqaddum Air Base is located in central Iraq, in Habbaniyah, and is also known as Camp Taqaddum.

truth, he established a ministerial training school in London prior to World War I. A year after the outbreak of the war, he suspended the operations of the school to volunteer as a YMCA chaplain ministering to troops in Cairo, Egypt. He had an incredibly effective ministry to troops there, particularly among the Australian and New Zealand soldiers who participated in the Battle of Gallipoli in Turkey. Skeptics in the military and the YMCA doubted that anyone was interested in hearing this chaplain speak, yet hundreds of soldiers packed into the hut where he taught as he challenged them to rely upon God for all of their needs, both physical and spiritual. After his death, a collection of his devotionals was compiled into the book *My Utmost for His Highest*. Through this devotional classic, Chambers became my daily mentor while ministering to Marines and sailors in combat. His devotional directed readers to fully surrender to God, not considering the consequences. As we do so, we can trust God to meet all the needs of those who are influenced by our decision to surrender to Him. I thought of Rosslyn and Eliza and all that they were experiencing due to my decision to obey God and become an active-duty chaplain. This was going to stretch my family, but we as a family also believed that God could not only get us through this, but make us stronger.

On Friday morning, 12 March, I and members of my unit collected the remaining pieces of our gear. After lunch, a convoy of 7-ton trucks—accompanied by our battalion's Mobile Assault Company in ar-Ramadi, 25 miles west of the air base—came to escort us to our new home for the duration of the deployment. Our first stop was Camp Combat Outpost, the home of two rifle companies on the eastern edge of ar-Ramadi. It was, let's say, "rustic." There were a few real, solid buildings in various stages of dilapidation, but the 300 Marines were actively making it feel like home. Our convoy was greeted by the resident company commanders—Captain Christopher J. Bronzi of Golf Company and Captain Kelly D. Royer of Echo Company. They welcomed us to ar-Ramadi with sly grins on their faces, knowing looks that seemed to say, "We're going to have an interesting time during this deployment."

After dropping off the Echo Company Marines, the rest of us made our way to Camp Hurricane Point on the western edge of the city, a former presidential compound wedged between the Euphrates River and a canal used to divert water to Lake Habbaniyah. Compared to Combat Outpost, Hurricane Point was lavish. The main palace on the estate had been destroyed, presumably during the initial invasion of 2003. There were, however, still some buildings that remained, including a large ornate building consisting of three sections connected by covered walkways. The battalion headquarters was located in the center section with the H&S Company headquarters and the battalion aid station (medical) on either side. The Navy corpsmen assigned to the 2d Battalion, 4th Marines, had set up the battalion aid station and claimed a room for Plaisted and me. We got settled into our new accommodations, which were nicer than we had anticipated. After our first night in ar-Ramadi, we received word on Saturday that four Marines had been wounded. It did not take long for the reality of combat to settle in. Plaisted and I caught a convoy back to Combat Outpost as soon as we could to see if we could offer any support to the platoon that had been hit. I am not sure that I knew what I was going to do when I got there. For most of the day, I felt like I did not do much; I simply talked with Marines and corpsmen. Eventually, I would learn that my presence was enough, but I was not yet convinced.

We stayed there over night, and I led a worship service on Sunday morning. We had a good-size group show up—about 40 Marines. During the sermon, I asked the participants how many hoped to go home more spiritually mature and strong. The vast majority of those gathered willingly raised their hands. I sensed the awesome responsibility of being a guide along that journey. If they were going to go home better and stronger men by the end of this deployment, they would need to battle cynicism, fear, hatred, and anger. They would be pushed to evaluate everything they believed in and question their own motivations while holding desperately to the lifelines of friendship and love. By the end of the day, I felt emotionally, physically, and spiritually drained. Ministry in this setting would draw upon all my resources, which were limited. I would need to make sure that I stayed close to Jesus—just as Father Devine had recommended at Camp Pendleton in January.

On Monday, 22 March, at 0315, the XO knocked on my door awakening me. The dreaded moment arrived—we had experienced our first combat death. Major Wylie came to me as soon as he received word. A rocket-propelled grenade killed Lance Corporal Andrew S. Dang along a road that unit members travelled frequently.[4] We all knew that we were in dangerous territory, but the reality of our vulnerability became incredibly real in that moment. I kept thinking about Rosslyn and Eliza. I imagined a car pulling up to my house in Temecula, with an officer and chaplain going to the door to break the news of my death. My heart broke at the thought of my family having to experience that. But I could not dwell on my vulnerability and my fears. I was there to be a chaplain, to be a symbol of hope, a sign of certainty even when everything else was falling apart. There was no time for the luxury of self-focus; I needed to support those around me. We had our first memorial service on Wednesday, 24 March. We held it in front of the battalion command post. Following my words and the remarks by battalion and company leaders, Marines filed past the memorial of boots, rifle, helmet, and dog tags to offer their final respects. Tears were shed by some; others maintained a steel-faced composure. But, it was obvious that everyone was thinking about death, justice, and survival, and I wanted to do my part to make sure that they had the resources to face each of those issues in constructive ways.

At the end of March, I was reflecting on where I was and what I was doing. A year earlier, I had felt like it was time for my family to leave our parish in Coquille, but I did not know where we would go. I never imagined that a year later I would be in Iraq with Marines, providing spiritual care for young men in combat. In fact, if someone had told any of my high school, college, or seminary classmates that I would someday be serving with Marines in combat as a chaplain, none of them would have believed it. I was not known for being the adventuresome type, not prone to taking substantial risks, and not known for hanging out with "rough" people. But,

---

[4] LCpl Dang of 1st Combat Engineer Battalion, 1st MarDiv, I MEF, was killed on 21 March 2004 while on patrol near ar-Ramadi. Associated Press, "Marine Lance Cpl Andrew S. Dang," *Military Times*, Honor the Fallen database.

here I was, and I was loving it, even though it was the most difficult thing I had ever pursued. It was a strange mixture of complete vulnerability and complete confidence. This was the place I needed to be.

Our second Marine was killed on 30 March.[5] I was there when we told his squad that he had been killed while on a walking patrol. His closest friend stormed out of the room in tears, and I followed him. He turned around and embraced me. He wailed with grief through his sobbing tears. His body was shaking as he held on to me for a very long time. I stood there like a sponge soaking up his pain, letting him express it all knowing that I would in turn give it to God. Receiving the pain of broken hearts was my job, and it was an incredible privilege to do so. But it did take a toll on me. On Saturday night, when I made my weekly telephone call home from the AT&T trailer, my own tears flowed as I shared with Rosslyn the events of the week. It was the first time that I had let my tears out since arriving in Iraq. And the next day, we had two more deaths.[6]

Early Monday morning, 5 April, Father Devine and Religious Program Specialist Third Class Edmond Garrett joined Plaisted and me on the convoy to Combat Outpost. Father Devine travelled a lot throughout the area of responsibility (AOR) since he was one of the few Catholic chaplains in Iraq. This was his first visit with us since we had left Kuwait. We had a good day with the Marines. When we arrived, we served breakfast to Echo and Golf Companies. Entering the metal building used as a dining facility, the Marines were expressionless but smiled when they saw us. We greeted them with warm words, kind smiles, and powdered eggs. Before Father Devine visited us, I had established the pattern of serving breakfast on Monday mornings when the logistics convoy brought hot food from

---

[5] LCpl William J. Wiscowiche of 1st Combat Engineer Battalion, 1st MarDiv, I MEF, was killed 30 March 2004 in al-Anbar Province. Associated Press, "Marine Lance Cpl William J. Wiscowiche," *Military Times,* Honor the Fallen database.

[6] Cpl Tyler R. Fey of 1st Combat Engineer Battalion, 1st MarDiv, I MEF, and PFC Geoffrey S. Morris of 2d Battalion, 4th Marines, 1st MarDiv, I MEF, were killed on 4 April 2004 by hostile fire in al-Anbar Province. Associated Press, "Marine Cpl. Tyler R. Fey," *Military Times*, Honor the Fallen database; and Associated Press, "Marine Pfc. Geoffery S. Morris," *Military Times*, Honor the Fallen database.

the U.S. Army base across town. Because of my predictable routine (something Father Devine had recommended), the Marines knew it was Monday because I was there serving breakfast. "Chaplain, is it Monday already?" they would ask in astonishment. Every day seemed the same, but my regular visits were visible signs that time was actually passing by.

I usually went back to Hurricane Point on Tuesday evening with the afternoon convoy, but Father Devine and Garrett needed to get back early, so we headed back on the Tuesday morning convoy. By the time we got close to Hurricane Point, we gathered that something bad had happened. When we got to the battalion command post, we went into the combat operations center (COC) and learned that a squad of Marines from Echo Company had been ambushed just east of Combat Outpost. That was the beginning of a series of events that resulted in our worst day in Iraq. By that evening, 12 men were dead, including the platoon commander whose Marine committed suicide in Kuwait.[7] Father Devine and I spent the day bouncing between my office and the COC. When we were outside, we saw smoke rising over the city, and we knew that our Marines were in the thick of intense fighting. We paused often to pray and to share our concerns with each other. Things eventually quieted down, and Lieutenant Colonel Kennedy, who had been out in the middle of the fray all day, came back to Hurricane Point. The three of us sat down in the ballroom that served as our battalion conference room. Lieutenant Colonel Kennedy started telling stories of the incredible bravery and heroism demonstrated by the Marines throughout the day. It was undoubtedly the most intense day of fighting that most of the men had ever experienced, and yet they revealed their true characters when the bullets started flying. Kennedy was justifiably proud of his men, and it was a privilege to hear him recount the tales of courage.

On Wednesday morning, 7 April, the four of us headed back out to Combat Outpost. Echo and Golf Companies had sustained the most casualties, and we felt that those Marines needed our support more than anyone

---

[7] On 6 April 2004, 12 Marines of 2d Battalion, 4th Marines, died in an urban firefight with insurgents. LtCol Kenneth W. Estes (Ret), *U.S. Marines in Iraq, 2004–2005: Into the Fray* (Washington, DC: Marine Corps History Division, 2011).

else in the battalion. Father Devine cancelled his other travel arrangements, and we made our way through the eerily quiet streets. When we got to Combat Outpost, everything was very calm. The Marines were exhausted from the previous day's fighting, and they were still trying to process their thoughts that 12 men had been killed. We made our way through the various places where the Marines were living, checking on their states of mind and offering encouragement when necessary.

Early the next morning, the entire battalion conducted a citywide sweep. I wandered into our dining facility at dawn and observed a bizarre sight. Cereal bowls and milk boxes were left on tables, some half-eaten. It appeared as though Marines had been eating and then left suddenly. Later, I realized that security for the FOB had been left to several dozen Marine cooks and Navy corpsmen. I believed those Marines could handle it. I knew that force protection was the primary job for the corpsmen, and I recognized that the FOB had a large perimeter with a fairly low wall. When I pulled all of that together, I suddenly became anxious about my own personal safety. I calmed myself by acknowledging that Father Devine did not seem particularly concerned. An hour later, Father Devine found me in the BAS. "Brian, I just realized that we're pretty vulnerable right now. What do you think?" I agreed with him and then went to my rack and started praying. About 30 minutes later, a second lieutenant from the Truck Company walked into the BAS, calling out, "Where are the chaplains? I'm taking them back to Hurricane Point."

The Marines of Truck Company had driven a convoy to Combat Outpost with hot food for the Marines when they got back from the sweep. Battalion leadership also realized that both their chaplain and the division chaplain were waiting at the vulnerable FOB. Someone decided that we should probably be taken to a safer location. Needless to say, I was incredibly relieved. I packed up my belongings and headed out to the trucks. As I walked out of the BAS, the corpsmen lined up along with Lieutenant Son. Suddenly, I was overwhelmed with emotion. I was being whisked away to safety, but these men had to stay. They were just as vulnerable as I had been minutes before. Would they survive? Would there be an attack on the

FOB? I said my farewells and hugged them. When I gave my farewell to Lieutenant Son, my tears started to flow. He assured me that the Marines would be okay and told me to get out to the convoy. One Marine died that day, but not at Combat Outpost.

Sunday morning dawned, and it was Easter. On 11 April 2004, Plaisted and I had a short sunrise service at Hurricane Point before we went to Combat Outpost for a memorial service. A normal Easter Sunday would consist of worship services filled with joyful songs celebrating the Resurrection. This was not a normal Sunday morning. We would get to the celebration of the Resurrection, but first we had to acknowledge the pain caused by death. Sixteen men died that week—15 Marines and 1 Navy corpsman. We did our best to honor them that Sunday morning, digging deeply into our spiritual and emotional resources so we could keep pressing on. There was no other option. After the memorial service, we went back to Hurricane Point. Two Marines had requested baptism, and so we concluded our Easter worship with the sacrament of holy baptism. A shallow pit was dug in the grove of trees beyond the main building, and then the pit was lined with heavy plastic and filled with water. The symbolism of being buried with Christ and raised to new life was particularly vivid as onlookers actually saw the baptized Marines come up out of a pit that looked much like a shallow grave. It was a fitting way to conclude a day of mixed emotions.

The battalion continued experiencing death throughout the deployment, but never on the scale of that first week, Holy Week, in April 2004. We held memorial services for everyone who was killed, but we started having them every two to three weeks instead of after each single death. We had grown weary, and it was difficult for everyone when we had a memorial service followed by another several days later. At each event, I would craft new words, but I used the same theme of recommitment: after such a tragic loss, we must recommit ourselves to the mission, to each other, and to God. I noticed that I was busiest after dark. When the sun went down, Marines wanted to talk. They wanted to discuss the pain they had experienced, the whys about what they were doing, and their relationships with God and others. I developed a deep bond with Father Devine through our times together as we laughed, prayed, and cried. I had learned to put my full confidence in God, especially on the days when I did not think I could get out of my rack and face the pain that I found in the lives of the Marines.

Sergeant Major Booker and I worked together to help Marines who were feeling overwhelmed by their experiences. We pulled individuals back to Hurricane Point and gave them several days to collect their thoughts and get a sense of renewal. I gave them writing assignments, and then we discussed them. Sergeant Major Booker sat on the back terrace and talked with the young Marines for hours. I am convinced that it was this focused time with the sergeant major that helped the Marines regain a sense of their identity and initiative.

At one point toward the end of summer, I was in Lieutenant Colonel Kennedy's office. Kennedy was sitting at his desk working, and he looked up at me and said, "Brian, you were exactly the right chaplain for this deployment. And you know that nothing will ever compare to this experience. You will compare every assignment to this one." I had a strange sense that he was absolutely correct.

By the end of the first week of September, we were getting ready to turn over authority to the next battalion, 2d Battalion, 5th Marines. As companies were turned over to Marines from 2d Battalion, 5th Marines, we went to Camp Junction City, the large U.S. Army base southwest of ar-Ramadi. Plaisted and I went over with the first wave of Marines to prepare for the warrior transition briefs (WTB). Before returning home, everyone was required to go through the WTB. It was a fairly basic concept. Help the men think through their experiences, crafting narratives that provide some structure to their time away. Then, help them understand what will be different when they return, how to cope with it, and what to do when they realize it is time to get help. There was specific guidance on how this was to be done, and it was a participatory exercise. I prepared to do this brief with every platoon as they arrived at Junction City. Over the course of 10 days, I conducted 30 sessions, each lasting 90 minutes. It was intense. The Marines had been through so much, and they knew that things would be challenging when they got home. The heaviest emotional time in each brief was when we acknowledged those friends who had been killed. Invariably, there were tears as the men were confronted again with the terrible cost of freedom.

Finally, on Wednesday, 22 September 2004, the battalion convoyed to al-Asad Air Base and flew to Kuwait. We were back at Camp Victory. We left Kuwait on Saturday, 25 September. Our first stop in the United States was in Bangor, Maine. It was an early morning fuel stop at about 0300, and we got off the plane to stretch our legs. As we made our way into the terminal, crowds of ordinary American citizens lined the corridor cheering for us. They would stop us to shake our hands and greet us. An elderly lady with white hair stopped me, looked into my eyes, and said, "Thank you," and then proceeded to put her arms around my neck and hug me. I was overcome with tears, and to this day, more than nine years later, I still get teary-eyed when I think about it. It was a practical expression of gratitude for what we had experienced from someone who seemed to appreciate the horrible price that had been paid. It had been 217 days since Clari took me to Camp Pendleton.

BEING HOME

Rosslyn, Eliza, Clari, and Sue were at Camp Pendleton to greet me. I cannot describe the joy, the gratitude, and the love I experienced at that moment. Eliza had learned how to walk while I was away, and she seemed fascinated by all the commotion in her serious sort of way. I hugged my girls and did not want to let go. I woke up on Sunday morning, 26 September, in Temecula. Since my family had given up our home in Oregon, we needed a new home, and Rosslyn and Eliza moved into our new house the day after I left for Iraq. The first thing I noticed was how quiet it was at home. The Marines at the battalion took a month of leave, and I enjoyed every moment I had to reacclimatize to the surreal world of peaceful suburban existence. We travelled a bit, explored our new community, and celebrated all that God had done for us. The month went by much too quickly, and I returned to work in late October.

Everything had changed. Some Marines had transferred. Every company had a different training schedule. During the deployment, I had access to everyone in the battalion. I knew what my role was, and I had felt necessary. Now, everyone was scattered. There were men I never saw again. Everyone was reestablishing support networks. I felt unnecessary. Of course, I was glad to be home with my family, but I was experiencing existential boredom, and that was very confusing to me. I felt guilty for wanting more satisfaction in my work, but I certainly did not want to go back to Iraq any time soon.

There was one significant task on my plate: I had to coordinate another memorial service. The battalion had conducted services in Iraq for everyone who had been killed, but now we were going to have one for every fallen Marine in the unit and allow family members to attend. It was my task to not only design the service but contact the families of all 34 men who had been killed. During the course of a week in early November, I called each family, praying each time before I picked up the telephone. It was an emotionally draining experience, once again exposing myself to the pain of those suffering. And I do not think I handled that exposure as well at home as I did while in Iraq. When I was deployed, I knew that the absence of normalcy required additional emotional and spiritual resources. While at home, I had a false sense of stability, and I did not access the resources necessary to cope with the pain.

During one evening at dinner in early December, Rosslyn asked me, "Is this PTSD?"[8] I was at the table, but it seemed like I was a million miles away. No, I did not have PTSD, but I certainly was dealing with the aftereffects of experiencing war—both the pain and the sense of professional satisfaction, which now seemed absent. Thankfully, Father Devine realized that the chaplains from the 1st Marine Division, who had deployed to Iraq, needed time to process their experiences. On 15 December, he sponsored a day of reflection at a monastery in Oceanside, California. A senior chaplain from another command led us through a series of conversations that provided opportunities for us to express and own our emotions. It was a powerful time of sharing and resulted in some relief.

By February 2005, I was reassigned to another battalion in 1st Marine Division. It was standard practice to shift chaplains after a deployment. My

---

[8] Post-traumatic stress disorder.

new unit, 3d Assault Amphibian Battalion, did not deploy as a battalion. We sent out platoons to Iraq, and my experiences with the 2d Battalion, 4th Marines, proved invaluable as I did preparatory work with the Marines and their families. Additionally, I worked with the Marines and their families upon the platoon's return from Iraq. Because I had been there with an infantry battalion, my guidance had some authority. Then, in July 2006, my family and I left California and moved to Norfolk, Virginia, where I was assigned to the USS *Vella Gulf* (CG 72). Life and ministry with Marines had become familiar to me, and I was very comfortable in that role. Now, I was going to the "blue side," working with the Navy for the first time. It took some recalibration as I faced new kinds of stressors and a different culture. And Lieutenant Colonel Kennedy was correct; I compared everything to my time with the Magnificent Bastards. I had a deep sense of solemn gratitude as we left for the East Coast—gratitude for the selflessness of the Marines who gave everything to their nation and to their fellow Marines and gratitude to those who shaped me as a person and a chaplain.

If I had words of encouragement for the men and women of the Marine Corps after reflecting on these experiences, I would say that Marines should never give up the culture of concern for the individual that they have developed over the centuries. Having served with other branches of the military, I know that no one understands care for the individual like the Marine Corps. I pray that the Corps will never lose that distinction.

REFLECTION POINTS

- Some Marines and soldiers suffer crises of faith. How does having a chaplain in the unit help those who need support and guidance?

- Chaplain Weigelt mentions how the core values and traditions differ among the various branches of the military. What values did the chaplain witness? Can you compare how basic values in your life or from your service have helped you?

- As policy makers adjust budgets to meet the needs of a nation, what funds need to be allocated to the military? What kinds of programs do servicemen and -women need during and after deployments to balance the experiences of war with life at home?

- Chaplain Weigelt's use of the words *we*, *they*, and *our* throughout the essay is a reflection of Weigelt's own connection to the battalion as a cohesive unit. Why would it be important for a servicemember and unit leadership to recognize that mentality before, during, and after a combat deployment?

CHAPTER SIX

TRUST AMONG . . . ALLIES?
STAFF SERGEANT MICHAEL R. MOYER

Staff Sergeant Michael R. Moyer has served nine years on active duty in the U.S. Marine Corps infantry. He deployed in support of OIF and OEF. He served with the 1st Battalion, 2d Marines, the Marine Corps Security Cooperation Group, and the School of Infantry-East as a combat instructor with the Infantry Training Battalion and will serve with the 1st Battalion, 7th Marines. Moyer was meritoriously promoted to staff sergeant on 2 January 2016. The following story focuses on his time as a squad leader with 1st Battalion, 2d Marines, in Afghanistan in 2010, how the Marine Corps' training prepared him, some of the challenges of his deployment, and the transition home with his Marines.

I grew up hearing stories about the U.S. Army in World War II and the Marine Corps in Korea, and then, while I was in the seventh grade, on 11 September 2001, the United States was attacked. My father was supposed to be at a meeting at the World Trade Center complex when the towers were hit, but he had decided to stay home to watch me play a sports game that afternoon. I felt that my family and I had a lot to be thankful for and that I owed it to the country and those who went abroad to defend the nation. At age 17, between my junior and senior years of high school, I enlisted in the Marine Corps, initially under a Reserve contract with the intent of going to college. As my senior year progressed, I realized that I did not know what I wanted to do with my life, and I was not about to put myself into debt paying for college when the plan for my future was unclear. I switched my contract from Reserve to active duty and then left for Marine Corps Recruit Depot Parris Island, South Carolina, two days after graduating from high school in 2007.

In 2008, I deployed as a lance corporal to al-Anbar Province, Iraq, a place known as the Sunni Death Triangle.[1] While my Marine unit at the time (1st Platoon, Charlie Company, 1st Battalion, 2d Marines) sustained no casualties, many of us witnessed some of the horrors of war—young men given suicide vests and sent to Iraqi Army or police bases and checkpoints. While in Iraq, I worked extensively with both Iraqi Army and police on raids, quick reaction missions, and interdiction operations catching smugglers near the Syrian border.[2] My experiences during this deployment proved to be extremely valuable when, a few years later, I worked to train my Marines and then led a squad through combat in the Helmand Province, Afghanistan, in 2010.

For one year, from March 2009 to March 2010, the 1st Battalion, 2d Marines, at MCB Camp Lejeune, North Carolina, conducted predeployment training for Afghanistan. At that time, the standard rotation of personnel and leaders continued to occur. What remained constant throughout this period, however, was anticipation of a rough deployment. Knowing that the enemy was already battle tested and waiting only stressed how important it was to develop well-trained and extremely disciplined Marines. To accom-

---

[1] Insurgency activities operated throughout cities within the Sunni Triangle, which was located from ar-Ramadi to Baghdad to Mosul. Estes, *U.S. Marines in Iraq, 2004–2005.*
[2] International news coverage at the time mentioned issues with smugglers of foreign fighters into Iraq along the Syrian border.

plish this, the unit's training focused on brilliance in the basics, a concept that originally did not seem high speed or sexy enough, but proved to be vital for the success of our mission and to minimize Marine casualties.

Our preparation for the deployment in Afghanistan began with a battalion-level team leaders course for the young Marines, many who would become squad leaders. The focus of this course was nothing more than getting back to the basics of the orders process, defense, patrolling, squad organic weapons, offensive fundamentals, and sound field craft.[3] The course culminated with a one-week defensive training operation including IEDs, ambushes, reconnaissance patrols, security patrols, 9-line medevac requests, and call for fire training.[4] After two weeks of written evaluations, performance-based evaluations, and a conditioning hike, each Marine was sent back to his company and platoon well prepared for the battalion's first major field operation.

Charlie Company, my company, dug in a defensive position that we maintained for a week while conducting patrolling operations. The company had to make sure that all Marines understood the priorities of operating in the defense and would conduct themselves accordingly. Digging proper fighting positions, performing tactical resupplies, patrolling, operating listening and observation posts, and utilizing proper radio procedures were stressed at every level. Up to this point, we had only heard of "training to a standard," but none of us had ever been rewarded for achieving that standard. When our position was inspected by the battalion commander and found to be satisfactory, 7-ton trucks were sent out to us, and we were taken back to the barracks. After that experience anytime the standard was not met, we all knew that we had to reset and hit it again until we got it right.

Tactical decision-making games were a significant component of our training for both squad and team leaders. Nearly 50 percent of Charlie Company was young as far as leadership and experience were concerned.

---

[3] *Organic weapons* refer to those assigned to the unit, such as an M16 rifle, a 40mm M203 grenade launcher, a squad automatic weapon, and K-bar or bayonet knives.

[4] A 9-line casualty medevac request is a casualty reporting method whereby Marines relay nine points of information for a medical evacuation.

In 1st Platoon, we only had three noncommissioned officers when we deployed. To balance for this lack of experienced Marines, we had to develop extremely knowledgeable and competent individuals at every level. In addition, in many cases team leaders were rack mates with members of their teams during recruit training, and so extra emphasis was placed on training the team leaders tactically and developing their abilities to make tough decisions under pressure. Our objective was to enable the young Marines in leadership billets to stand out from their peers.

In October 2009, the battalion went to Marine Corps Air Station (MCAS) Yuma, Arizona, to conduct a monthlong training exercise. The U.S. Army Yuma Proving Ground is the home of an advanced aviation school. At Yuma, we were able to focus on everything from basic rifleman tasks, such as proper employment of grenades, to company-level heliborne night raids. In Yuma, the company could train at the lowest level of buddy rush and work all the way through company raids.[5] Squad leaders used white-space training to show their Marines the effects of terrain on different formations.[6] Platoons were able to work internally training on fire and maneuver, and as a result, each Marine developed a firm understanding of final coordination lines and what the weapons attachments could add that the platoon did not already have.[7] When thinking about the logistics and the many challenges that take place while conducting a company-level night raid that uses helicopters, there is a lot that can go wrong. Company leaders, however, did an excellent job of breaking down each part of the order and the mission. By the time the order was briefed to team leaders, it was simple for the Marines to understand "go here, move here, shoot here, and make sure that the correct Marines are near me." This large-scale raid, when broken down

---

[5] Buddy rush is a strategic movement where one Marine provides cover fire, enabling a second Marine to run forward before stopping and providing similar coverage to allow the first Marine to move forward as well.

[6] *White-space training* refers to the use of realistic training areas, scenarios, weapons, and equipment that a unit will fight with.

[7] Fire and maneuver is a tactic that uses suppressive fire to increase a unit's ability to move forward by decreasing an enemy's ability to return fire. The final coordination line is used to coordinate the shifting of support fire and movement as troops move into a final formation just prior to a unit's assault on a target.

properly, was as simple as "shoot, move, communicate," a philosophy that all infantrymen are familiar with. Our brilliance at the basics motto was evident in the performance evaluation checklists that were handed out from the company to appropriate leaders. The Marines were equally evaluated, while all leaders were fully aware of the training standards to be achieved. In Yuma, the Marines learned each other's strengths and weaknesses and how to work together.

The next large exercise was held at Marine Corps Air Ground Combat Center Twentynine Palms, California. Exercise Enhanced Mojave Viper was the culminating event for the battalion prior to the deployment to Afghanistan. The purpose of Mojave Viper is to evaluate a battalion as a whole from squad-level tasks up to company-level live-fire evolutions. While at Twentynine Palms, we received additional training on Afghan culture. We were briefed on *Pashtunwali*, the lifestyle of Afghan people, and we were also given the opportunity to interact with several Afghans who live in the United States.[8] The cultural education, however, was rather general and broad and did not delve into the specifics of Pashtun or other tribal lifestyles. At Twentynine Palms, we conducted several live-fire exercises, which were the apex of all we had trained for, including platoon assaults, company assaults, mechanized operations and attacks, and patrol base operations at both company and platoon levels. Most of our training focused on assaults and attacks, how to react to an IED, and other fairly clear-cut scenarios.

After the Mojave Viper exercise, we knew there was not much more we would do prior to deploying. Our weapons platoon sergeant went to a course to become proficient in the process of requesting and utilizing air support in a combat environment. Upon his return, he immediately began working with the company squad leaders to give Marines a solid understanding of how close air support (CAS) works. It was a much-needed training because many of us had never spoken to pilots or called in air support in combat. The close air support training was helpful because of all the

---

[8] Pashtunwali is a way of life and system of customary laws that stress honor above all else. See Maj David W. Kummer, comp., *U.S. Marines in Afghanistan, 2001–2009: Anthology and Annotated Bibliography* (Quantico, VA: Marine Corps History Division, 2015).

different terminologies used when speaking with pilots. However, besides CAS training, we were mainly taking care of paperwork, updating our wills, checking on everyone's emergency contacts, and submitting our predeployment leave requests.

In March 2010, as a lance corporal, I deployed with the battalion to Helmand Province.[9] Musa Qal'ah was the district where our battalion set up its Combat Operations Center. Our area of operation (AO) was mainly desert with canals used by the locals to irrigate their fields. Less than half a mile from the irrigated fields, vegetation ended and the land turned into a flat desert. In the desert, there were a few sparse small villages or towns, established around wells that provided water to irrigate the surrounding fields. This environment was rather different from Iraq. Towns and villages were smaller; the fields farmed by the locals were also considerably smaller than in Iraq. The al-Anbar Province in Iraq was more technologically advanced than Helmand Province, Afghanistan. There were plenty of paved roads in Iraq, while I did not see even one gravel road in Afghanistan; the only exception were the roads on our bases built by the Coalition forces.[10] Electricity was generally unavailable in Afghanistan, and an entire village took turns irrigating fields from a single canal or stream, while the Iraqis I worked with could rely on the TharThar Canal and Euphrates River for water.

In Afghanistan, the battalion's mission was to conduct counterinsurgency (COIN) operations. We did this by befriending the local populace, working with and training the Afghan National Security Force (ANSF), and neutralizing the enemy either by detaining them or by, when necessary, killing them. Engaging members of the local population, getting to know them, and talking with them was something that many of us Marines had already experienced in Iraq. We found that the people in Afghanistan were tired of the Coalition's unfulfilled promises and ideas. To address and change this negative attitude, the battalion used Afghan money and workers

---

[9] Moyer received a combat meritorious promotion to corporal in May 2010 during his deployment to Iraq.

[10] The International Security Assistance Force (ISAF) in Afghanistan, commonly referred to as the Coalition, at this time was under the direction of the North Atlantic Treaty Organization (NATO).

to boost the economy in the villages and towns, helping everyone with projects, such as mosque renovations, electrical line repairs, and establishment of new water wells.

The members of the ANSF that we worked with were Pashtun, and many of them were either from the same tribe or from an equally respected and powerful tribe.[11] I remember one Afghan Army lieutenant who was motivated, eager to learn, and enthusiastic about working with us, but who was not helpful because he was of a different or less powerful tribe than all of his soldiers, and so he was not listened to. The majority of the ANSF troops we worked with were from the local area. This had both positive and negative implications. On the positive side, they knew the area extremely well and were invested in the safety and success of the battalion's missions. On the negative side, occasionally they would lead a patrol and just go to their families' homes and sit and drink tea instead of patrolling the AO or moving to areas that lacked protection from the battalion's patrols and posts. We conducted a few joint operations with soldiers of the ANSF in which they took on leadership roles while U.S. Marines only provided support. The intent of these operations was to increase the ANSF confidence and experience in leading missions while building and improving relations between them and the local people. The outcome of these operations was mixed. Often the soldiers did not appear to be very effective, while others were great and made a positive impact on both the ANSF and the locals.

Working with the ANSF, both Afghan National Army (ANA) and Afghan National Police (ANP), was not easy and presented many challenges. It happened that some ANSF members stole from Marines. Most of the ANSF members smoked marijuana and showed up for post or patrol high. At one point, I informed the Afghan Army senior enlisted that his soldiers had been growing weed in our landing zone. Following that incident, we had fewer cases of intoxicated ANSF on post or patrol, but when it did happen, the Marines were extremely uneasy about having ANSF with weapons anywhere near us.

---

[11] Pashtun is a tribal or ethnic group located mostly in the mountainous regions of eastern and southern Afghanistan and in Pakistan.

Punctuality was definitely not a priority to the ANSF troops, and that was extremely frustrating to all of us. Marines would get up after only three or four hours of sleep so that they were ready and on time for patrol, only to sit around for an additional 20 minutes, on a good day, waiting for members of the ANA or ANP. We were willing to trade 20 minutes of sleep in exchange for a meal or even mail from home on some days, so to waste it sitting around waiting was very frustrating and contributed to bad feelings between Marines and members of ANSF.

Not all ANSF interactions were poor, and every opinion will vary from Marine to Marine even within the same AO. I was fortunate enough to have been friends with an ANP commander who was quick to the fight, eager to keep his AO safe for both Afghans and Coalition forces, and extremely helpful to work with. He shared intelligence and local rumors with us and supported us in any way possible. His example inspired the policemen in his unit, other ANP units, and the locals of that village. He was so effective in his operations and working with the Marines that the Taliban targeted and killed him while he was traveling to the patrol base (PB) where I had been operating out of for a meeting. Fortunately, his replacement was just as enthusiastic and committed as he was and picked up right where the previous commander left off. I remember standing on the roof of the compound these commanders ran, analyzing their security to see if they had missed any avenues of approach, and they pointed out their homes to me. This proximity to home was the center of gravity for each of these commanders. They were close to home, and they wanted to raise their children in a safe and good environment. This is why they were so successful, yet this was also the reason why the Taliban targeted them.

One evening while standing sergeant of the guard with my Marines on post, I heard a single shot fired from our southern entry control point.[12] I immediately called down to ECP South to ask what was happening and was informed that the ANA just started shooting. I found our interpreter and went to ask the ANA why their soldiers were shooting. The response I

---

[12] Sergeant of the guard is the supervising noncommissioned officer (NCO) of enlisted servicemembers pulling guard duty.

received astounded me: "The post to our south is not answering their radio. So we shoot to wake them up." At the time, this was both frightening and hilarious all at once. The fact that the Afghan soldiers might have been sleeping on post did not seem to concern them, while the most efficient way to wake the soldiers, instead of patrolling or driving out to them, was simply to shoot in their direction. Perhaps some of the humor came from sleep deprivation, but at the time, keeping a straight face while explaining to the ANA soldiers that perhaps that was not the best idea was very difficult.

What I found truly disturbing in Afghanistan, even more than in Iraq, was the complete disregard for human life displayed by the Taliban. In Iraq, it was not uncommon to hear a Marine telling his junior servicemembers that if children were around, you knew you would be okay. While conducting the relief in place with the next battalion in Afghanistan, I had to warn the newly arrived Marines that they would get killed if they relied on that belief.[13] Open areas where children played, houses, fields, and roads or bridges were all fair game for direct fire and IEDs. I remember responding to an IED strike where the Taliban had targeted the ANA because they set their checkpoints up in the same spot every day and did not have electronic countermeasures (ECMs) like the Marine battalion did. Unfortunately, on that day, a child was running up to the ANA soldier when the Taliban detonated the IED. It was not a pressure plate or timer but a remote-controlled IED, which means that someone was watching as that child approached the soldier and made a conscious decision to detonate at that time.

This callousness struck even the hardest and coldest of us Marines, even those who had been in battles with large numbers of casualties. I remember talking with a fellow Marine, a good friend of mine, who was extremely upset about the situation. He just looked at me and said, "Nobody wants to see dead kids, man. F——g nobody." After treating small children or keeping our ECP open so that families could bury their dead, every man was ready and itching to find the Taliban and kill them, but the inspiration and anger we felt those days were not worth the weight of the sights, sounds, and smells that we were left with.

---

[13] Relief in place is when all or part of a unit is replaced by an incoming unit.

Toward the middle of the deployment, the relationship between Marines and ANA became very strained. Several things played into the stress felt by both sides. While on post, an Afghan soldier and Marine got into a fist-fight over mere trash talk, but both sides felt slighted and were not entirely happy even after the issue was resolved between the two individuals. Additionally, the ANA senior enlisted soldier, affectionately known as the "Sergeant Major" had been killed by an IED. This was hard particularly for the ANA but also for the Marines, because if the Marines had a problem with an Afghan soldier, we could always go to the Sergeant Major, and he would take care of it. He was a good soldier and one of the best ANSF leaders I worked with. The Sergeant Major was a middleman between the ANA and the Marines, though he did it without compromising his position as a leader. His soldiers did not look at him as merely the mouth of the Marines or their commanding officer; they looked to him as the guy who had been there and done that and respected him. He had the ability to take what we, as Marines, were asking of him or trying to teach him, and adapt it to the Afghan way of doing things, in a manner that got nearly the same results. After his death, one ANA soldier tried to kill himself with rat poison. Following that event, we were briefed that some of the other soldiers may attempt "death by Marine" where they would point their weapons at us so that we would kill them. Such a possibility added more stress to an already tense relation.

Corruption was a problem among the Afghan security force while the loyalty of some of its members was questionable. In a number of cases, ANSF members were caught or killed while emplacing IEDs or conducting attacks on Coalition forces. These types of scenarios were not uncommon and truly biased all of us Marines toward ANSF prior to setting foot in Afghanistan, which made the following event a difficult one. One morning, at 0300, my squad and I began a patrol to establish a listening and observation post (LP/OP) near a route that we knew was used by the Taliban. We were in place by about 0345 and settled in for what we hoped would be a quiet night. We displaced to return to base (RTB) at about 0500, just before the sun rose.[14] On the way back, the day transitioned into dawn, that hour when

---

[14] RTB is an order, either displayed or verbal, to proceed to another point or site.

everything is gray but visibility is good and getting better, so we were able to lift our NVGs up, which was important for spotting small IED indicators on the ground. While walking on a tree branch acting as a bridge over a canal and exiting the vegetated zone of our AO, we took contact. Most of the squad had made it across the bridge already, but we had a team that had not crossed yet. Automatic-weapon fire hit in between the second and third fireteam from a compound to the southeast of our squad. We immediately dove into the canal for cover and oriented ourselves in the direction the fire was coming from.

When I looked up at the compound, I noticed a flag flying over it and a square structure on the corner, very similar to the posts we built on our PB manned by the ANSF. When under fire, many thoughts go through a person's mind. It is impossible to describe the fear you experience thinking that friends, the men you are responsible for, might be hurt or killed. It is the kind of thing you cannot imagine until you have lived through it. In addition, some people experience outrage and a desire to respond in a forceful way to those attempting to kill them, especially when the shooters are supposed to be allies. I found myself in a dilemma. Do we return fire and enter the compound? Or, do I stop this before it becomes a massacre of ANSF and an international incident with my squad and I at the center of an investigation?

We had received *only* one burst of fire, and I was able to establish that none of my Marines were injured. I immediately called for my Marines to hold their fire. I then called out to the post that we were Marines. After a few seconds of strained silence waiting for a reply, we got a "sorry" from the ANA soldier standing post who merely put his hands up and shrugged his shoulders. Naturally, my Marines were upset and started yelling at the ANA and making their way toward the compound as if they were about to make entry. I quickly got their attention and tasked my team leaders out and got everyone on the road again. Once in the PB, the squad was extremely frustrated and wondered why we get shot at and why we did not go into the compound. Before I could explain myself, however, we had a rotation on post from 0600 to 1200. I told them that we would sit and talk through it after post. When our rotation was over, I explained to them everything that I had been thinking and my reasoning for not going in. After several cigarettes, some grumbling, and some more explaining to ensure that everyone understood the reasoning instead of just hearing it, we were all in agreement and carried on with our day.

That is just one example of the stresses of combat. The doubt, fear, confusion, lack of intelligence on the enemy, or knowledge about the safety of friends is what we call the "fog of war." At times, it is thick and we find the simplest tasks to be difficult, and at other times, we feel as if we are merely away from home for a while. Dealing with all these factors that make a major impact on us as human beings—while still trying to locate, close with, and destroy an enemy—is a significant task. Additionally, each Marine reacts to combat differently, so all Marines have to know where their buddies are and pay attention to ensure that the Marines are physically and mentally okay while they are still doing their jobs. Throw into the mix a group of undisciplined, possibly disloyal "allies," and combat can become overwhelming. Those feelings, doubts, and emotions take a toll on those in battle, and as so many veterans and their families have seen, the consequences do not always present themselves immediately.

A few months after the shooting incident, our unit left Afghanistan. Returning home from that environment was exhilarating. It was like finally having a weight lifted from my chest. At the same time, however, it was frustrating and awkward. People changed and so did the homes and lives that we left when we deployed. For Marines on deployment, it is as if life at home is just put on pause, and we expect to see it the way it was when we return. Accepting those changes is just part of the transition of coming home, realizing that everyone kept moving and their lives continued and changed. For some, it feels as if everything is moving in fast-forward while trying to catch up on all the changes that took place. Many Marines become frustrated from trying to keep up with everything.

Most servicemembers spend a great deal of time thinking about their homecoming. When we loaded up the trucks for the last patrol or boarded the airplane for the trip home, it felt like being on the top of a roller coaster about to take a huge drop. All the excitement and emotions that we had bottled up for the last several months were all present, and we felt like bursting.

The most defining moment for me was when my Marines flew out, but I had to stay behind for one additional week to conduct the relief in place. One of our explosive ordnance disposal Marines, who had been working with me throughout the deployment, looked at me and said, "Mike, you look so relaxed." I was so relieved that my squad of Marines was safe and on their way home that I felt as if I were done. It was the most significant catharsis I have ever felt in my life. Of course, I still had patrols to conduct, and I was still in danger, but it felt as if that hardly mattered because my squad was safe.

For me, getting home was more about adjusting than just feeling relieved. How was I supposed to deal with the fear of waking up and not being able to find my rifle or not knowing where the guys I spent the last seven months with were? Most of us just did everything we could to relax and adjust to being with family again. We had to become accustomed again to someone touching us constantly when wives and girlfriends wanted to hug and hold hands. Patience was the best tool for the families and the Marines coming home—patience in understanding that it might not feel right for a little while and we might have to make some sacrifices on both sides to make it work.

When you are taken to a place where there is no running water or electricity, often not enough or no good food and little chance to get a good night's sleep, the day-to-day things you are used to no longer matter. Which new phone or music is coming out this month? Who is dating or getting married? Often Marines get angry because we do not care about the little problems that people will complain about not realizing how nice their lives really are. Go up to someone who has never deployed and ask him or her to imagine seven months without a day off and no reliable communications with home. The reactions are usually along the lines of "Why?" But, that is the reality for us coming home. The real struggle is letting go of the bitterness we feel for others simply because "they haven't done it." Wives with children did not get a day off, and I have yet to meet a mother or father who did not worry while a child was deployed. And it is nobody else's fault that we enlisted in the Marine Corps during a war and then went to war! All of this reasoning, however, is hard to accept for someone who has seen friends be killed and had to pick up their body parts.

In addition to the adjustment from deployment to home life, Marines must adjust again as their friends, leaders, and juniors move on either in the Marine Corps or in their lives. Many Marines will move to other duty stations or be sent to another battalion while others will get out of the Corps and move back home. New Marines arrive at the battalion, and everyone must start the cycle of preparing for a deployment again. Restarting this cycle is particularly frustrating due to the high level of proficiency most squads gain toward the end of a deployment. The Marines begin again at the lowest level of training, and even that does not work smoothly due to the new members in the unit.

Many Marines feel anxiety about returning home. The routine and cycle of life that we lived with every day for seven months or longer change drastically. The personal issues that had been put aside for the past several months reappear again, in addition to any new concerns that may be brought home from deployment. Anxiety about moving to a new area of the country, starting a new job with a different command, or taking on extra responsibility in a higher billet are all in the back of many Marines' minds.

When I returned home, I felt lost, out of place, and as if I did not belong there. I had been in charge of a squad for seven months, and then, when I came home, those Marines were no longer around. Instead of living with my squad of Marines 24/7, I saw them for a couple hours, and then we went our separate ways for the night. Even when I was with my wife or staying at a friend's house, I felt alone. While my Marines and my friends took their postdeployment leave, I checked out of the battalion and moved to Virginia. For 30 days, I had no responsibility for anyone except myself, and I did not know how to spend my time or what to do with myself. That emptiness was akin to boredom and restlessness. Not knowing how to spend free time or what to do with oneself is why a lot of Marines fall into dangerous habits, such as alcohol abuse, reckless driving, or drug use. Spending time with friends and family was what kept me going and really allowed me to start to move forward with my life.

Isolation, confusion, loneliness, and feelings of being lost are not the only things felt. There was the excitement of seeing loved ones again or sharing the joy of a new child in the family with your friends. Many of us who served in different platoons and did not see each other much over the deployment later got together to drink and laugh about the different things that happened. Many of our friends who did not deploy with us, or got sent home early, were waiting for us with coolers of beer and chicken to throw on the grill eager for everyone to catch up on what was missed both at home and abroad. For many of us, the support we received from family and friends was extremely important, yet the support we received from fellow Marines—who picked us up when they saw us falling after we came home or just drank with us in silence and then made us get up and do something or drove us home to our families—made a significant difference. The Marines who take care of each other after getting home are the ones who have a successful transition home; those are the guys that veterans stay in touch with.

These are just examples of the bond Marines have with each other. Anyone who has ever been there knows the feeling of being at the bottom of the barrel and then coming to terms with the situation. A buddy will say, "Now stop crying, p——y," and both of you have a laugh about it. It is not a disparaging comment, not when it is coming from the guy who just sat with you until you worked out whatever was bothering you. Or maybe, he is the guy you do not want to let down, so you try to suck it up until he pulls you aside and asks what is going on. The bond we Marines have is not about shared tattoos or drinking stories; it is that level of comfort we have because we have seen people at their very best and very worst, when they have looked like they could not do anymore, and then we have looked around and realized everyone else is in the same boat, so we keep pushing on. Together, we have tested our limits morally, physically, and emotionally. We have seen the entire spectrum of each other and that is a closeness that some married couples do not even experience. The stress of our work and our feelings and the pain that comes along with our job is the foundation of a strong bond. It is the mortar that holds us together and the reason the Marine Corps is known for brotherhood and esprit de corps.

Looking back, I have realized that I did not make much progress with the ANSF members who I worked with. I did not teach them much, and I do not feel that they contributed much to our success on deployment. The next time I deploy, regardless of the time or location, I will spend some time focusing on working with members of a foreign military—understanding how to teach them and how the rapport I will have with them will affect our mission. The Marine Corps spent money on instruction for cultural differences, provided COIN classes, and sent us on operations that focused almost entirely on the local populace of Afghanistan. If training time had been spent with the host nation counterparts and Marines had been given more time to focus on that aspect of COIN, our operations would have been quicker and the success of those operations more substantial. The United States is in the process of turning over Afghanistan to the Afghans, and that transition as well as the success of the ANSF would be more defined if every Marine battalion had focused on that mission instead of leaving this to the military and police transition teams.

More than just providing the classes and getting the information out, military leaders need to instill these mission values in their subordinates. Just like showing a PowerPoint presentation on alcohol abuse will not keep anyone from drinking too much, slamming information into Marines with a fire hose will not stick. Marines who understand and are invested in the mission need to be identified and given responsibility. If all of the Marines are mature and invested in the mission, anything will be possible.

To the Marines preparing to deploy, I would say, "It's on you." The time is there for training; it is on individuals and the leadership to think of and schedule events and to always put out and get the most out of every training evolution. Leaders need to connect with their Marines, make them want to work for us. Ask any enlisted Marine which was more effective, the a——s chewing or the calm talk about disappointing performance, and they will tell you that the calm talk made them feel worse and want to work harder. I believe that there are two things we cannot teach, no matter how good we are as an organization. We cannot teach leadership, and we cannot teach someone to care. If we connect with each other, if we bring brotherhood into the picture during training, then we could bring more Marines home.

REFLECTION POINTS

- Not all Marines or other personnel experience PTSD, but all have to come to terms with returning home and the changes that come with that return. Did you go through this process? Was your experience more like the chaplain's experience or like Staff Sergeant Moyer's?

- Both Lieutenant Colonel Beyer and Staff Sergeant Moyer mention the brotherhood felt by Marines. How do you define the Marine Corps brotherhood? Do their definitions differ in concept or just expression?

- Trust is an underlying theme of Staff Sergeant Moyer's recollections. How did the Marines' lack of trust toward the Afghan soldiers potentially damage the battalion's mission? How could leaders have addressed these concerns?

CHAPTER SEVEN

A WOMAN IN CHARGE: A CIVIL AFFAIRS MARINE TEAM LEADER EXPERIENCE IN AFGHANISTAN
MAJOR ANIELA K. SZYMANSKI, USMCR

Aniela K. Szymanski is a major in the U.S. Marine Corps Reserve and a judge advocate/civil affairs officer. Currently assistant counsel at Judge Advocate Division, Civil and Administrative Law Branch, she was the rule of law advisor for 2d Civil Affairs Group, and served as Civil Affairs Detachment commander leading four civil affairs teams. From February to September 2011, she was a civil affairs team leader attached to 3d Battalion, 2d Marines, as part of Regimental Combat Team 8 in Now Zad District, Helmand Province, Afghanistan. In this chapter, Major Szymanski shares the challenges she faced as a team leader during that deployment.

. . . AND HOW GENDER IS NOT ALWAYS WHAT YOU MAKE OF IT; SOMETIMES IT IS MORE

When I decided to join the Marine Corps in 1999, my mother thought I should be committed to a mental institution. She envisioned her little girl being either eaten alive by ruthless male Marines who would never have her best interests in mind or that I would transform into a grotesque hybrid of a man and woman. In the end, she supported my decision and continued to be my biggest fan. Even beyond my perhaps over-protective mother, some acquaintances and friends reacted very negatively to my choice. The best way I can characterize their general reactions is "Why would you ever want to do that?!" as if they could not understand any possible reason why a college-educated woman would make such a choice. Some attempted to rationalize my decision in their own minds by saying, "Well, the Marine Corps is paying for your law school, so that's why you're doing it." When I corrected them that the Marine Corps did not pay for even one cent of my law school, they too were left utterly confused. For those who have not experienced it themselves, the call to serve and the desire to be a Marine is really beyond explanation. For that reason, I stopped trying to explain my choices to people.

The Marine Corps proved to be the most professional and respectful organization in which I have ever served. I am, by no means, a man-woman hybrid, and every Marine who I have ever had the pleasure of serving with has been exceedingly honorable in their actions and intentions. Throughout my Marine Corps career, my gender has never been an issue, that I am aware of, until my deployment to Afghanistan.

I served seven years on active duty as a judge advocate after completing OCS and TBS. I eventually realized that I would need to establish myself as a civilian attorney-practitioner if I wanted to have a career outside of the Marine Corps, and decided to take a civilian job but stay in the Reserves as a major. As a part of the Reserves, I joined the 4th Civil Affairs Group (CAG) in Washington, DC, and was assigned as a civil affairs team leader. I knew nothing of civil affairs prior to joining the unit, but was told that my legal experience would be an asset. I quickly learned that civil affairs actually had nothing to do with law and everything to do with human relations. In the Marine Corps, the term *civil affairs* covers such a large variety of civil-military operations, including displaced persons operations, humanitarian assistance, and relations with foreign national governments. Some of these functions are not even interrelated, but for the fact that they all have the common denominator of, in some way, dealing with civilians.

My detachment was assigned to deploy in support of OEF in 2011. I was tasked by my detachment commander with assembling a team of 11 Marines who would be responsible for assisting an infantry battalion in Helmand Province, Afghanistan. Our deployment took place as part of the troop surge about which senior political and strategic leaders had a precise idea of its purpose. At the tactical level, our level of operations, the ISAF intended for us to support Afghan forces, to play a leading role, and to take responsibility for their nation, their people, their security, and their governance. I knew that, if this was going to happen, we were going to be the ones to do it. For that reason, I also knew that I needed the best team I could assemble. I decided to set some specific metrics about the type of Marines I was looking for.

I determined that the qualities I was looking for in a Marine on my team would include good physical fitness (judged by their PFT and CFT scores), credibility within an infantry battalion (based on their past assignments, service, accomplishments, and occupational specialties), and some level of accomplishment in their careers (meritorious promotion, selection for special duty assignments, college degree).[1] Team members also needed to demonstrate their willingness to go above and beyond the call of duty, and be individuals who expressed some level of humbleness in their communication skills. This last factor ended up being the most disqualifying among the Marines I considered. I would not compromise on this point. I firmly believed that I needed Marines who could be both solid team members to their fellow Marines and be compassionate to local civilians who had already endured nearly a decade of war with the United States and a lifetime of war overall. Our mission's success depended entirely on their willingness to give of themselves for the sake of others whom they had never met and to be considered trustworthy. For anyone, this would be an extreme challenge; for Marines going into a war zone, this would be a rare gem. I also had my team chief, Gunnery Sergeant Earl Beatty, and my as-

---

[1] PFT stands for Physical Fitness Test; and CFT for Combat Fitness Test.

sistant team chief, Staff Sergeant Charles Frangis, review all of the potential team members' Reserve qualification summaries to rank them. Their choices included Corporals Cody Banks, Travis Neal, and T. J. McCabe, all of whom were promoted to sergeant during our deployment. This ensured that my team chief and assistant team chief had some ownership in the process and the team. Besides that, I trusted their judgment as to who would be able to handle the demands of our mission.

I ended up with a team of all men, almost all infantry Marines, except for my assistant team leader, Captain Gregory Hudgins, who was a logistics specialist and had participated in the initial invasion of Iraq in 2003; a corpsman, Hospitalman Second Class Parker; and a communications specialist, Lance Corporal Jamal Lea. Only my team chief had performed a civil affairs mission previously, and none of us had been to Afghanistan yet. Thankfully, the Marine Corps had approximately a decade of lessons learned from Iraq and Afghanistan for us to study prior to deploying. For several months before our deployment, we studied after action reports for civil affairs relevant material; spoke with Marines who had recently returned from conducting civil affairs missions in Helmand Province; attended seminars and programs at the U.S. Institute of Peace in Washington, DC, and with other organizations about topics relevant to establishing economic and government stability in Afghanistan; and performed map studies of the region.

We spent every drill weekend conducting some scenario-based training exercises, which turned out to be an excellent investment of our time. These exercises consisted of having Afghan role players and interpreters set up scenarios that Marines may encounter in Afghanistan (e.g., communicating with *shuras*, encountering farmers on patrol, interacting with local small business owners), and using an interpreter to communicate with Afghans.[2] Mainly, the training scenarios were designed to get Marines familiar with using the District Stability Framework (DSF), and comfortable communicating through an interpreter.[3] As painstaking as this was for the Marines,

---

[2] Shuras are meetings of the village elders.

[3] The DSF is an assessment and management process to support improved stability operations in an area.

myself included, the level of comfort that the Marines felt once arriving in Afghanistan and using these skills, the repetitiveness and tedium that they felt during the training was well worth it.

Approximately two months prior to leaving for Afghanistan, my detachment commander informed me that there was an issue with me being the team leader. He said that the battalion commander was concerned that my gender could potentially jeopardize the mission. Basically, the concern was that with the sex-biased culture in Helmand Province, the local Afghan government officials and tribal elders would refuse to deal with me. The current battalion commander in charge of the area of operations had convinced him of this. The regimental commander supported the battalion commander's position. My detachment commander said that he would do everything he could to ensure I was able to remain the team leader, but that he was not sure what could be done.

Naturally, this caused a great amount of stress for me personally, and uncertainty for our team. I simultaneously prepared my assistant team leader to take over as the team leader, and conducted as much research as I could to determine whether the battalion commander's concerns were legitimate. I wrote an e-mail message to a civil affairs team leader who had been in the district I was to be assigned to, and he stated that the Afghan governor had little respect for women and only saw them as sex objects. I spoke with two Afghan cultural advisors working for the Marine Corps, and they, in contrast, vehemently disagreed with the belief that a woman could not be successful as a civil affairs team leader in Afghanistan. They explained that the Afghans would have great respect for a woman and that I would likely be granted access to areas where men could not go because their culture requires them to be deferential to women. They did not agree at all that elders or government officials would refuse to work with me. I explained all of this to my detachment commander in support of my bid to remain as the leader of my team.

My detachment commander arrived in Afghanistan more than a month before my team. He called me on the phone and explained to me that he had met with the battalion commander and regimental commander, and finally with the general in charge of Regional Command-Southwest for ISAF. He told them that if they wanted a civil affairs team in the district, it would have to be with me as the team leader. If they refused to take the team with me as the leader then they would not be receiving a civil affairs team in that district at all. I was deeply touched by his confidence in me and by his support for me remaining in my position.

When I arrived at Camp Leatherneck, Afghanistan, a few days in advance of my team, I was still uncertain about whether the battalion commander would coalesce with my detachment commander's position. I was directed to report to the battalion commander, who was in the neighboring district to the one I was to be assigned to. I met with him and the operations officer, and they informed me of their thoughts on a woman leading a civil affairs team in their area of operations. They explained quite bluntly that their belief was that the Afghans would treat me with disrespect and perceive me as a slut or prostitute. They doubted the Afghan district governor would take me seriously or listen to my advice or guidance. My perception of the meeting was that, at best, they doubted that, as a woman, I could ever be successful and, at worst, my gender was jeopardizing the success of the entire COIN operation in the area.

This would have been shocking to me had I not been aware of their thinking about a woman leading the civil affairs team. Thankfully, I had been able to prepare myself. I explained that, if at any point during the mission I, or anyone else, felt that my gender was creating an undue risk to the success of the mission, I would have no problem stepping aside and allowing my assistant team leader to take over. This apparently reassured the commander enough to allow me to proceed to the district I was supporting, with my team following a few days later.

We were assigned to the district of Now Zad (sometimes spelled Nawzad). It is a district consisting of a rural farming area at the southern foot of the Hindu Kush mountain range. It has a beautiful vista of the Hindu Kush and flat, desert landscape. Areas around the *wadi*, or underground river, were green with crops. At some point, UNICEF had been present to assist in schooling children, and USAID had been in the area to build wells for the local communities. An organization had also built a clinic in the district center. I was told that, in days long gone, this was the vacation spot for Helmand Province. There was a hotel, restaurants, and gardens. It became the rest and relaxation spot for insurgent fighters.

Fierce fighting had taken place in Now Zad from 2006 to 2009, first with British and then with U.S. forces. On 4 December 2009, Marines launched a decisive offensive dubbed Operation Cobra's Anger that lasted for three days. The fighting was so intense that Marines who were there at the time nicknamed the district "Apocalypse Now Zad." In fact, a friend of mine, who was a Cobra pilot and participated in Cobra's Anger, could not believe that there was anything left to go back to by the time it was my turn to go there. He was correct. The bazaar in the district center was absolutely obliterated. Partial brick walls remained in some portions of the bazaar, and twisted steel heaps scarred the remainder of it. The clinic had been shot so many times that there was a bullet hole every 8–12 inches throughout the entire building, except where a large portion had been blown away. The mud and clay houses that people used as their homes, along with the mud and clay walls around their compounds lay in ruins. The entire local population had fled as a result of the intense fighting during Operation Cobra's Anger. By the time my team and I arrived, some individuals had returned to their homes and farms in an attempt to reestablish their lives. The governor had been installed and was working out of the former clinic, as it was being reconstructed part by part.

After completing the turn over with the civil affairs team in the district we were about to replace, I was able to establish a manpower distribution that could support the lines of operation within the mission and as part of the broader Helmand plan. We fell in on some important reconstruction projects, including rebuilding the school that had been operational, rehabilitating the bazaar, and fixing the roads that had been damaged during the fighting. The previous team had begun the projects that were designed to support the return of economic stability to the region as well as restore freedom of movement to the local populace. Some locals, however, would never return because they had established new lives for themselves in other districts, while some were seeking to return to reclaim their properties and livelihoods.

The civil affairs Marines embedded themselves on every patrol they could. Their objective was to gather information to assist in identifying sources of instability in the district and areas where the Afghan government needed to improve if they were to be perceived as legitimate and responsive to the locals. The district had essentially never had representation from the central government, as it had always been a tribally governed area, as much of Helmand Province was. Now Zad was inhabited entirely by the Pashtun tribe, with the subtribes consisting of Alizai, Barakzai, and Noorzai. Attempting to get local buy-in to a government that was unfamiliar to them was extremely challenging.

The infantry battalion concentrated on establishing security. Their training was focused on fighting the enemy. That is what combat troops do, of course. The battle for Now Zad occurred in phases. I arrived when the district center was relatively secure, and the battalion was trying to recalibrate to working with the local population and the ineffective government. Yet, most of the district was in the hands of the Taliban and, with them, those tribes who remained hostile to a centralized district government.

Civil affairs tasks concentrated on attempting to resolve disputes between local civilians, the Marines, and their own government, while trying to lower the tensions following major destruction caused by combat operations. My predecessor ran a very expensive "cash for work" project, paying men to perform unskilled labor such as filling in dirt roads to make them passable in order to show a positive role of the Marines. My challenge was to develop the local governance. There was a district governor who was regarded by the battalion as too corrupt to trust with some legitimate cause. They did, however, have tea with him daily, meanwhile controlling the district by force.

The civil affairs team had access to funding for local projects and this helped to create some leverage with the governor. I ran this relationship. I think the governor and his small number of officials found it puzzling to have a woman in this role. The battalion saw civil affairs as keeping peace with the locals, but not being part of the overall mission.

The center of Now Zad was mostly secure and, given our role and the nature of civil affairs, we wanted to improve our relations with the locals.

Some were living just outside the gate of our forward operating base. We met with them at shuras, and they often came to the gate to ask for help, which often involved cash. Patrols went out from the base, always heavily armed and in big vehicles. Nothing could be less friendly to people who had lived their lives in those sandy lanes with unsupervised children running through the streets and who had ran small shops along the roads.

So, the civil affairs team tried to engage the locals just outside our base and during patrols. Initially, we walked out within sight of the observation posts on the base with just the basics—a uniform, a weapon, and a radio—just as the team previous to us had done. The locals found us approachable and, without the protective gear, we seemed to make them less fearful of imminent attacks. Yet, battalion commanders changed their views and ordered that full protective gear be worn under all circumstances for personnel leaving the base. The locals were puzzled and did not act as friendly toward us, in some cases even refusing to shake our hands as we walked through the town. For me, it was disappointing that the battalion commander did not prioritize the effect on the local trust in making such a decision.

The battalion provided excellent security for the district center and succeeded in expanding the security area significantly during the seven months that my team was there. This allowed the Afghan government representatives the opportunity to expand their influence by reaching out to those disenfranchised areas while demonstrating the government's usefulness and responsiveness. The extent to which they were willing to do so, however, was limited initially. In order to accomplish this, it was necessary to get government officials in contact with the people. When I arrived in the district, the governor refused to leave his compound to go out into the community. His position was that if the people wanted something from the government, then they should come to him. Although he did not say so, he seemed also to be concerned about his security outside his compound—numerous times he attempted to convince me that he should be given a Marine Corps-issued pistol to carry for his own protection. Knowing this was no way to convince the locals that the government was on their side, I persuaded the governor to come out and meet the locals with me in shuras outside of the district center and during walks through the bazaar. This is one time when my gender was an asset. Among other methods I attempted, I basically ended up shaming the governor into leaving his compound by telling him that if I could go out among the people, and I am a woman, then surely he could as well. I personally accompanied the governor to shuras with elders as well as during walks through the district center to check the progress of the bazaar reconstruction and other projects. I expressed my support for the government and its initiatives, and encouraged locals to bring their problems to the government so that the governor could address them.

My civil affairs Marines located in different areas of the district were able to feed information back to me about what was going on where they were, how the government was being perceived and how effective the ANP and ANA were. I was then able to use the information to develop plans of action for the government. With the support of the battalion's security, we could address the security needs of the locals, their concerns about employment opportunities, and problems caused by a significant level of corruption. In 2011, the population of the district increased dramatically. Although a census was never performed in the district, the number of shops in the bazaar increased from about 10 to more than 30, and the estimated number of occupied residences in the security area increased from about 25 percent to about 50 percent. The locals with whom I spoke stated that security, which used to be their primary concern, was now so good that they felt strongly encouraged to return to their homes.

After security, the locals' next concern was the high level of unemployment. The Afghan people we spoke with during population engagement efforts said that the lack of jobs was their main nonsecurity-related problem. Farming was the major source of revenue in the district, yet local farmers could not grow the crops they needed to be profitable because of the lack of water. *Karezes* (ditches) that fed water to the crops were largely damaged or destroyed during Cobra's Anger because insurgents used them to create a tunnel system to smuggle weapons and fighters. Without employment opportunities, some locals took jobs from the insurgents from whom they were tasked to build and emplace IEDs. The ANP in the district did not have enough manpower to establish a regular enough presence to identify these activities before they cost lives. The civil affairs team worked with the ANP advisor, a Marine captain, and developed a plan that, if a community provided a certain number of recruits for the ANP, the civil affairs team would build a well in the community. This plan resulted in only limited success, as some families did not want their sons joining the ANP because of its tarnished reputation of an abusive organization.

Without enough water to grow crops that yielded a sufficient profit, farmers were easily persuaded by the insurgency to grow poppy. The insurgents paid the farmers a fee up front, provided them with a water pump and generator, and furnished the other supplies needed to grow the poppy. When the poppy was harvested, the farmers received the remainder of the fee and the insurgents took back the water pump and generator. Marines saw huge poppy fields during patrols and convoys, but were prohibited from disturbing them. We were instructed that poppy eradication was a Drug Enforcement Agency (DEA) role and that Marines would be disciplined if we invaded its jurisdiction. Attempting to convince these farmers to change from growing poppy to rice was almost akin to a comedy hour. The district in which we had seen an attempt at a poppy eradication program once during our deployment was an unmitigated failure. During my seven months in the district, I did not once see a DEA representative. Two fields were partially plowed before the plows were destroyed when the Afghan government attempted a poppy eradication program effort, and the ANP basically gave up within a matter of days. This likely had something to do with the fact that the ANP commander was growing poppy.

As was the case everywhere in Afghanistan, corruption by Afghan government officials was pervasive. My civil affairs team was able to assess through our population engagement that the local government's corruption had become a major source of instability. For the citizens of the district, corruption of the local government was indeed a major problem. The district stability team and the battalion commander were able to use various means to encourage local government officials to adjust their behavior, thereby leading to a change of the perception of the local people. This happened through close mentoring by the civilian British stabilization advisor, Hans Swift, sent forward by the Provisional Reconstruction Team. In a very short time, the public perception began to improve.

Clearly, civil affairs was a vital tool in the success of transitioning control to the Afghan government and would have even been a greater force with effective interagency coordination. A COIN operation is most successful when it is "firing on all cylinders." Based on what I observed in Now Zad, as security was being established, other operations should have been moving forward more aggressively than they were. Resources for locals to reestablish their agricultural livelihoods should have been immediately offered and put in place. Unfortunately, all resources that we tried to access had a long delay in delivery or could not be delivered at all. A program that provided solar-powered water pumps for wells to allow farmers to pump out water from their wells to irrigate crops was "not available in Helmand" because of "security concerns." A wheat seed distribution program took more than six months to deliver the seed to the district. The civil affairs team was prohibited from buying wheat seed from a local source to distribute because that would compete with the formal seed distribution program that had been established. Our assistance to local farmers was basically limited to handing out shovels in exchange for attending a shura with the governor. In the meantime, insurgents were providing local farmers with free generator-powered water pumps, fuel to grow poppy, and payment in advance for the harvest. Essentially, we could not compete with what they had to offer in the agricultural livelihood respect. Regretfully, this resulted in critical lost momentum.

The Marine Corps was able to be adaptable and responsive to the situation within the district to get to our desired security end state; but without the same responsiveness and effort from other agencies, COIN was stymied. The insurgents clearly knew they could take advantage of this weakness, and they did so. Because of this, battalion commanders must decide on one course of action to follow through on COIN missions. Either they should focus more effort on coordinated operations with other agencies or they should give civil affairs leaders greater latitude and authority to communicate directly with these agencies and tap their resources. The Marine Corps has stated that civil-military operations are "owned by" battalion commanders because it is their mission within their battlespace. Yet, often the Marine Corps is perceivably lacking in the way it supports civil affairs. Since the Vietnam War, the Marine Corps has known that civil-military operations will be an integral part of our future, but it has placed extremely low emphasis on this in its training and education of commanders, emphasizing instead other metrics of success. Regretfully such an approach is reflected in how poorly some battalion commanders view civil affairs. For example, our battalion commander had a daily, evening briefing. The brief had 60 slides, and many were the same every day. Out of the 60 slides, civil affairs was allotted one. I remember one briefing when the commander said a working dog was going back to the United States because of stress and then skipped over the civil affairs slide, as a result he failed to address any issue related to civil-military operations.

Marines are effective at what they do. Yet, the transition from battle to peace remains a challenge. Allowing peacemakers into the battle spaces is uncomfortable. There is, however, no battle that cannot benefit from the remaining local civilians to create or resume some form of self-governance. As a general principle, the importance of having local support during COIN operations cannot be overstated. Once any foreign military force arrives "uninvited" into a country, the dynamic is against it and it is likely that the invaded has sided with the insurgents. The Marines who come in first, who have to fight the bloodiest of battles, are battle hardened and are not going to deeply engage in civil-military operations. Their replacements likely will not either, as they are conducting security operations to maintain the momentum of the battle and retain hard-fought gains. So, it is at about the third rotation when the fighters start to think about the future and peace. That is when I arrived in Now Zad.

The district was a textbook COIN "success," as we transitioned governance back to the Afghans, complete with a functional police and army, while displaced people were returning to their homes. Yet, it could have been a greater success. If longer lasting success is to be had, the Marine

Corps must place more emphasis on civil-military operations driving the planning process at various stages, rather than the other way around. In addition, the interagency partners must deliver the outcomes of their missions as well.

In my role as a civil affairs team leader, I was still a woman. At all of the shuras I spoke at and in each encounter I had with locals, my gender never seemed to be a problem for them, and indeed, at times it proved to be a great advantage. Elders approached me, shook my hand, and asked me for the things their communities needed (e.g., schools, mosques, and jobs) just as they would have with a man. I sat next to the governor during all official meetings and events to show my support for him and his government, and I was never excluded from any function.

As a woman, I had access to information from local women and children that would have been inaccessible to a male leader. When I visited the local clinic, which was where women of the community congregated, I was mobbed by women who told me how happy they were that I was there and explained to me what their husbands, brothers, and sons needed to bring stability to their lives, which not surprisingly was jobs. There I also found out that some families had moved back into the district center because the security had improved. I learned of this from an enthusiastic young lady who also wanted me to kiss her baby. Children approached me on patrols, called me their sister, and told me about their day in school. This information allowed me to determine whether the teachers we were supporting were indeed performing their duties as outlined by the Provisional Reconstruction Team partnership with the Afghan Ministry of Education.

Even more fascinating, former Taliban fighters had no issue with my gender either. On one occasion, the Human Intelligence Exploitation Team had identified an insurgent willing to "reintegrate."[4] The Human Intelligence Exploitation Team, the psychological operations team, and I met with this individual as established by the local Afghan reintegration director. As part of my duties, I supported the reintegration program and offered any reintegration resources that may have been warranted based on the situation.

---

[4] A formal process of leaving the insurgency and rejoining mainstream society.

The man was approximately 50 years old and identified himself as a mid-level leader. I did not offer my hand for him to shake, instead just bowed slightly as I greeted him, but he did not appear to be concerned about my presence at all. The meeting lasted about 45 minutes and, although he was understandably nervous the entire time, I sat across from him without any issue whatsoever. So, it appeared, my gender was not even a cause for concern to insurgents who generally held extremist views about gender.

Beyond what I added to the Marine Corps mission, including being a woman leader, my team added a new dimension to the situations we encountered in Afghanistan. Naturally, I will never know the views of all the locals about me and my gender. I do, however, have some anecdotal stories of how my gender may have planted a kernel of thought among some of the locals. One of these was from an Afghan man who worked with the civil affairs team performing local construction projects. He had two wives, neither of whom left the home to work and they covered themselves in public. He had sold one of his daughters into marriage when she was only two years old, although the daughter continued to live with him until she was old enough to join her husband. This man told my interpreter that it was good that I was a woman in a position of power in a place like Now Zad. He said that maybe some of the Afghan men would see that their women can do things that men can do too.

Indeed, what initially seemed to be a liability proved, instead, to be an asset. I was able to communicate with the local women about their perceptions, needs, and challenges, and also with the local men about their women and children, something that male Marines would not have been able to do. I found the men were willing to speak with me about all subjects, but even more so about their own dreams and aspirations. A *mullah*, or religious leader, approached me for a microgrant to open a perfume shop in the bazaar and said that, because I am a woman, I can understand these "finer things." A young man who sought to rebuild a mosque destroyed during an ISAF bombing also explained to me that his sister, who lived in a neighboring village, could not travel to get medical care for some burns she suffered to her face and that the village was in need of a doctor to visit. As a result, civil affairs coordinated a visit to the area and organized a medical education program. The same young man, a military-aged male who could have been fighting in the insurgency, also came to me seeking assistance with the mosque restoration. His mother told him that he was to "work honest with the Marines" because we were "there to help." Both he and the mullah came to the gate of the forward operating base and requested to speak with me about their needs. I thought this demonstrated ease about my gender and a willingness to not let that stand in the way of progress.

My experience as a woman deployed with an infantry unit in a remote, isolated combat zone was very positive. The Marines at the forward operating base were highly respectful, always professional, and never acted inappropriately. My team, comprised of almost all former infantry Marines who had never worked *with* or even *for* a woman before, performed superbly and I never once had reason to doubt their loyalty to me or the mission. While the examples I have set forth are anecdotal, it is indeed these seemingly small shifts in attitudes and behaviors that play an extremely important part in COIN success. Reaching the tipping point entails the culmination of many individual efforts and events that yield a disproportionate change. Indeed, three years after the end of my deployment, I spoke with an interpreter who worked for me and had remained in Afghanistan. She shared with me that Afghan women who used to be afraid to leave their homes and always wore burkas in public, in 2014, felt safe enough to walk through district centers without burkas, only wearing *hijabs* (scarves covering their hair) instead. She also stressed how much more competent the ANA was and how well they were doing. The collective persistence and dedication of every U.S. servicemember and American civilian who served in Afghanistan has resulted in changes that will continue to impact Afghanistan forever.

There is no future conflict that will not involve civil-military operations. When the goal is defeating an insurgency, it is clear that one of the most important, probably the most important, centers of gravity in a COIN operation is the populace. And the Marine Corps would do well to use every resource at its disposal to gain the tactical advantage, including its women Marines.

REFLECTION POINTS

- Considering the other chapters in the book that were written by men, how is Major Szymanski's work and ethics different? How are they the same?

- The author has some pointed critiques about where the Marines and other American organizations failed to provide the support needed to secure the region. Does the weight of her criticism come from her position on a civil affairs team, a gender-based interpretation, or both?

- How do you feel about women in combat zones? Did this account sway your opinion in any way?

CHAPTER EIGHT

CROWD CONTROL
FIRST SERGEANT NICANOR A. GALVAN

First Sergeant Nicanor A. Galvan has served with the U.S. Marine Corps since March 1999. He deployed four times in support of OEF and OIF. His story takes place in 2003 when Galvan was a sergeant and infantry squad leader for India Company with the 3d Battalion, 5th Marines, during the OIF I invasion. Galvan focuses on how, with limited training, he worked with his Marines to unknowingly stumble through counterinsurgency operations. Galvan tells readers that there is nothing special about him, but he hopes this story serves to keep others from making the same mistakes he did.

Ignorant men raise questions that wise men answered a thousand years ago.

~Johann Wolfgang von Goethe

What you are about to read is my recollection of an event that occurred during my first combat experience while I was a squad leader with the 3d

Battalion, 5th Marines. I am a normal person, no one extraordinary. I am from a small suburb of Waco, Texas, called Elm Mott. I was born on Lackland Air Force Base in San Antonio, Texas, on 10 November 1979. My father is of Mexican-American ancestry, while my mother is Greek and was raised in Athens, Greece. Needless to say, I had an interesting upbringing. My father was in the U.S. Air Force and met my mother while stationed overseas in Greece. He retired from the Air Force in 1979, the year that I was born. I grew up going to military bases and dressing up in combat fatigues nearly every Halloween and watching the *G.I. Joe* cartoon. I would like to say G.I. Joe was the reason I joined the Marine Corps, but I think it goes deeper than that.

When I was growing up, I had this burning feeling inside like I wanted to fight or do something great. I just did not have direction or know what to do. It was not until I was almost 19 years old when I realized what that something might be. My friend, Trinity Dove, returned from Marine boot camp, standing tall and confident. I was intrigued by his transformation and all the fanfare he received from our mutual friends. I knew then that I wanted to change the direction of my life and do something great. I went to the Marine recruiter shortly after and was laughed out the door. See, I weighed 225 pounds at five feet five inches, which is not proportionate at all. The recruiter gave me some pamphlets and shooed me out the door, saying good luck and call when I have lost about 60 pounds.

Despite that, I still did not rush to lose the weight and began technical college that fall. By the winter quarter, I was tired of school and was determined to make a change. I woke up one morning in January and decided that I was going to lose the weight and join the Marine Corps. I ran and ate a can of tuna for lunch and a can of tuna for dinner every day for two months before I finally lost the weight. After reaching my goal, I walked into my recruiter's office and told him that I wanted to join the Marine Corps and be in the infantry. Shortly thereafter, I went to boot camp. Fast-forward about four years, and that is where this war story begins.

Prior to 11 September 2001, I had been in the Marine Corps for two years and was preparing to deploy with 3d Battalion, 5th Marines, to Okinawa, Japan, attached to the 31st MEU. I was a rifleman in India Company, and training was pretty normal. The unit trained in the hills and washes of the San Mateo area (often referred to by Marines as the "Backyard"), which is located on the north side of Camp Pendleton. We would train by doing buddy team rushes and going to the field for our MEU qualification exercises.[1] The mantra of the training was "brilliance in the basics," although it was not called that back then. We also had a unit organized training course of nonlethal exercises, which consisted of basic crowd control techniques. The training consisted of forming a phalanx, pushing a crowd back, and using snatch squads/teams.[2] This training was limited and probably only lasted about a week. I believe it is important to mention that no one in my platoon had any combat experience except for my platoon sergeant, who had seen combat during the Gulf War.

After the devastating terrorist attacks of 9/11, training did not change at all, and the battalion continued to prepare for the deployment to Okinawa. We deployed between January and February 2002 and executed a normal unit deployment program (UDP).[3] Upon return from this deployment in July 2002, Operation Enduring Freedom was well underway. The battalion began to break up with Marines executing permanent change of duty stations and Marines getting out of the Marine Corps. I, on the other hand, still had eight months before I was to leave the unit, so I continued to train with what remained of the skeleton battalion. The battalion maintained the same routine of training, did buddy rushes in the Backyard, and made a trip to Marine Corps Air Ground Combat Center Twentynine Palms, California, to conduct something termed by the battalion as Desert Palm exercise. All that meant was that each company in the battalion ran through the Range 400 series of exercises and then went home.[4] This training only lasted for a

---

[1] These exercises are part of a MEU's work-up period prior to deployments. The various training and exercises are intended to integrate the separate units to operate effectively together.

[2] *Snatch squad* refers to a crowd control tactic of rushing into a crowd of people with the intent of snatching one or more individuals who are attempting to control or lead a riot.

[3] The author's use of UDP is intended as a generic term for the unit's deployment to Okinawa at the time.

[4] Range 400 is a live-fire range at Twentynine Palms used to train rifle companies in the techniques and procedures for attacking fortified areas.

week to a week and half. From July to December 2002, this was the largest training event conducted prior to the unit getting the call in January 2003 to deploy to Iraq.

In December 2002, I, as a sergeant, had actually made the decision to get out of the Marine Corps but had a change of heart in the middle of the month just prior to going on Christmas leave. At that time, there was still no indication that I might deploy to Iraq. I decided one morning that I was not ready to be a regular civilian, and I wanted to remain in the Corps. That day, I immediately went to the career planner who drew up the paperwork, and I reenlisted. The plan was to go to the School of Infantry to be an instructor, and my orders were for February 2003. I went on leave and came back around 3 January. I picked up my checkout sheet that day and hurriedly got all of the signatures needed to leave the unit. While getting all the signatures, I remember rumors flying around the battalion that the unit was going to Iraq. I did not pay much attention, but when I saw my fellow India Company Marines getting tricolor desert cammies, I knew it had to be real. I had only one signature left to get before I was officially checked out of the unit and that was my first sergeant's signature. All the Marines in the company were at the battalion aid station getting their shots in preparation for the deployment. At that point, I did not know what to do. Was I going to volunteer to stay with the unit and go to war with the battalion or execute my orders? I found the first sergeant and said to him, "First sergeant, I have one signature left on my checkout sheet, and once you sign this, I am gone." I must have said it in a way that seemed like I wanted him to tell me to stay, but instead he asked, "Well, what do you want to do, Galvan?" I looked at him and said, "I will go." He immediately took me to the personnel office and had the Marines push my orders back as far as they could. Then, I immediately reported back to 1st Platoon, India Company.

As soon as I was back in 1st Platoon, I was issued my desert cammies and received my shots. Later that day, while back in the barracks, my staff sergeant called me into his office. He asked me, "Do you want to be the 3d Squad leader?" I immediately said, "Yes." I was really surprised. I had been a squad leader before, but I quit because the old platoon commander and I had not gotten along. It was different now; the staff sergeant and I got along well, and he was now the platoon commander. I had always told myself that if I ever had the opportunity to be a squad leader again, I would do it differently. Either that same day or the next, the platoon commander called all of the squad leaders together, and we had a draft to select our squads. At this point, the platoon, companies, and battalion were scrambling to prepare. The following week, the unit got a small drop of new joins from the School of Infantry, and I received four new Marines, which brought my squad total to nine. I had no time to waste training the squad because it was already the second week of January, and we would leave in early February 2003. Every morning, my squad was up at 0500 to conduct physical training. Because the days were so condensed with administrative preparation, there was hardly any time to schedule in training. I do not remember conducting Backyard training—just getting up early, doing physical training, and fitting classes in between inspections.

There was a lot going on, and I decided to make it even more complicated by proposing to my wonderful girlfriend, Beth, and marrying her just before leaving for war. The battalion then deployed in early February, went to Kuwait, and set up in a large tent city somewhere in the Kuwaiti desert. Once there, the Marines trained quite a bit. We often left the tent city and went to the middle of the desert in Kuwait to practice setting in the defense (e.g., digging fighting holes, setting sectors of fire, etc.). Usually, after setting in the defense, we practiced fire and movement and also conducted maneuvers with our amphibious assault vehicle attachments. This included some live-fire exercises. I remember doing our battlesight zeros and throwing live grenades. Still, the live-fire exercises were few and far between, and we mostly concentrated on practicing fire maneuver and fire movement, setting the defense, and integrating with our vehicle attachments.[5] We did, however, do extensive nuclear, biological, and chemical (NBC) warfare training. This consisted of spending hours

---

[5] For more information on the range of military missions that deployed infantry companies train for, see *Infantry Company Operations*, MCWP 3-11.1 (Washington, DC: Headquarters Marine Corps, 2014).

on MOPP training along with decontamination exercises. Lastly, we received limited language and culture training prior to executing the push to Bagdad, Iraq. Retrospectively, that training would have served the unit better in the later occupation stages. These types of training and maneuvers went on for about a month, and some Marines in my company, including myself, started to doubt if we would even go into Iraq. This was pretty naïve considering the amount of money that was spent getting so many military units to Kuwait.

Despite our doubts, our unit crossed into Iraq and began the Iraq invasion in mid-March. The invasion went well. We were constantly on the move, and my squad did an excellent job fighting our way to Bagdad, having one casualty. Looking back, I feel that my squad was trained well and prepared for the job required of it during the invasion. The mission was pretty simple. We drove in, set the defense, and if we took contact, we conducted our immediate action drill, returned fire, and moved on the enemy. This is exactly what we had trained to do. Unfortunately, the next challenge my squad faced had nothing to do with a majority of the training my squad had received leading up to the invasion.

Once our battalion finished its mission in Bagdad, we moved up to Samarra, Iraq, where we stayed for a short time before finally being ordered to move the battalion south of Baghdad, down to al-Diwaniyah, in early May. Al-Diwaniyah was a decent size city, and my squad was immediately thrust into action and placed in a bank. The position was turned over to me by four U.S. Army soldiers, who I believe were with the 82d Airborne Division. I was briefed by a sergeant first class, which I found peculiar as he was only in charge of three other soldiers. Despite this, I took over the tactical position and began to set up an urban defense, which I had little knowledge of since most of my training was for desert and jungle warfare. Luckily, the platoon guide assigned to me had a wealth of knowledge in urban training because of time spent with the Special Operations Training Group. My squad set up the defense and was told to secure the building from looters and to conduct security patrols around the area. Other than that, there was little to no guidance on exactly what our mission was or exactly what we had to do. Looking back on it, I feel like my company just improvised its entire effort in al-Diwaniyah and never really executed an organized plan.

By mid-May 2003, the unit's work became routine, yet the situation was more complicated than it seemed. Day-to-day operations in al-Diwaniyah consisted mostly of guard rotations, security patrols, and crowd control. The crowds consisted of children and Iraqi people who wanted to see what we were doing. In our minds, the war was over for the most part, but we still treated everyone like a threat even though the locals smiled and waved every time we were on patrol. I remember seeing a group of Marines in town without their gear, each carrying just a weapon while roaming the streets in buddy pairs. I later found out that those Marines were part of a light armored reconnaissance unit that allowed them to hang out in town for some liberty. That did not last long; after about two hours, the Marines were recalled back to their unit. This is a prime example of how confused everyone was about exactly what the Marines were supposed to be doing. One unit of Marines thought it was safe in town, and so the unit's Marines enjoyed some down time, while my unit believed that the local populace and situation could turn hostile at any moment.

Even though there was confusion about my unit's perspective on the state of the war, it was not long before the unit started to get direction as to what it needed to do. My squad was given a couple of Iraqi policemen to help guard the bank who were told that they would live downstairs in the bank, while my Marines lived upstairs. I found this scary because I was just getting shot at by the locals, and now I was supposed to trust these AK47-toting policemen. To complicate the situation further, the Iraqi bank staff had returned to work, cleaned the place up, and planned to reopen the bank. Our role started to change from fighting a war to aiding the Iraqi community to restart local services.

The following events that unfolded in the bank mirrored the change in my unit's role and how complicated the situation still was. I do not remember all of the details, but I believe my platoon commander told me that there would be some U.S. Army personnel coming to my bank to pay Iraqi teachers. One evening around the end of May, a Marine officer along with a couple of Army soldiers came to my bank to discuss the payment to these teachers. They brought thousands of American dollars and placed the money on a table. The Marine officer laid down my mission by simply stating that I was there to be security while the Army major and staff sergeant worked with the Iraqi bank personnel to pay the Iraqi teachers. From my understanding, this meant that I would have the usual security checks at the door, and the Iraqi teachers would be coming in and out. That night, I passed the word to the Marines in my squad, and everyone was tracking on what was going to happen in the morning. Nothing could have prepared me for what I was going to wake up to on the following day.

It was about 0800 when I woke up to what sounded like a huge crowd of people downstairs. I walked to the rail of the second deck where the Marines were located and peered down to the first deck where 100–150 people crowded together in a small foyer area screaming at the bank tellers, who were separated from the crowd by an extra-long wooden podium that the tellers had brought in days before. I immediately got my gear and told everyone who was not already downstairs to get their gear on and head down to the first deck. Once at the bottom, I saw the Army staff sergeant sitting at a desk near the counter, doing nothing. Then, I saw the Army major using my interpreter to try to calm the crowd down. I went straight to the major and asked him what was going on. He said that the people just rushed in once they opened the place. I pushed my way outside to the front of the building where I met up with two of my Marines on that post. I asked them why the hell they allowed all those people to come in, and they said the major told them to. I also asked why they did not wake me up when it got out of hand. The answers they gave did not make me happy, but then I noticed that there was also a significant crowd on the outside of the security perimeter set with concertina wire. The Iraqis outside of the perimeter were in the middle of the road blocking traffic.

Once I had the situation completely assessed, I went back inside to the major to ask him what he wanted us to do since he had caused this ridiculous situation. He said that since the crowd was being so unruly, that we should just kick them all out. Immediately, I got all the Marines in my squad that were not on post and formed a line between the counter and the staff sergeant's desk. I instructed my interpreter to tell the people to move back. The plan was to keep five Marines in a fairly tight phalanx and to push the crowd out the door one step at a time. I began to yell my commands. "Ready! Step! Ready! Step!" It was not long before the crowd started to squeeze out the door. Just as we began to gain momentum, the major yelled to me and said to stop and allow them to come back in. I was a little confused but complied, and the crowd immediately engulfed my Marines and me as we reopened the floodgates. Within another few minutes, the major found me and said to push all the people out again. I reorganized my small Spartan-like phalanx and began my commands again. "Ready! Step! Ready! Step!" Again, just as we began to make headway in pushing the unruly crowd out, the major again gave the order to let them back in. Again, I complied, and again, my Marines and I were engulfed by the Iraqi crowd. Minutes later, again, the major said to push the crowd out, and just like the two times before, we repeated the same process.

When engulfed by the crowd for the third time, one of my lance corporals turned to me frustrated and said, "Sergeant, these people are starting to not respect us because we keep allowing them to come back." I was already upset, but now that my Marine had said out loud what I knew, I was enraged and knew I had to take action. I turned around and jumped over the staff sergeant's desk and began screaming at him. "Look, I am tired of this. You need to stick to one plan, if you want them out then send them out. If you want them inside, sitting down in rows, then we can do that too." After this verbal tongue lashing from me, the major said, "That sounds fine." I did not wait to decipher which plan he wanted, so immediately I began to tell my Marines to get this crowd to sit down and shut up. The Marines followed my lead and along with the interpreter began to verbally and physically get everyone in the building to sit down. The Marines and I achieved this in about five minutes, and the Army soldiers were stunned. They could not believe how ferociously and quickly five Marines and an interpreter got a 100-plus-size crowd to sit down and shut up. Things were now calm inside, and the major tried along with the Iraqi bank staff to begin paying the teachers. On that day, the Army soldiers were only able to pay 30 teachers and planned to pay the rest the next day.

When the day was over, the major told me that I did a good job, but I made sure to tell him that this was not going to happen again. I stated that I would organize the security for the next day and direct who would come in and who would not. He had no problem with that. That evening, I sat with my Marines, and we devised the security plan for the next day. We decided that there would be two entrances to the bank on the security perimeter. One line of people would go into an alleyway along the wall of the bank, and the other line would form along the sidewalk on the other side of the bank. In addition, two Marines would stand watch outside the bank to ensure that the Iraqis stayed in line and no one crowded the street as they had the day before. My squad also planned to wake up early and be set in place before the crowd of teachers arrived. The following morning, we set in place and aggressively directed the Iraqi foot traffic. Throughout this whole process, we were, what might be considered, mean to the populace, but that is how we had been conditioned throughout all of our training. That day, the Army major and his staff sergeant were able to pay approximately 200 Iraqi teachers. This was an enormous difference from the previous day. The major left that evening thoroughly impressed by the tenacity of my Marines, and I was satisfied that we were able to improve on what had been a complete debacle the first day. I wish I could say that was the last time I dealt with crowds, but it was only one of many times when I would have to deal with crowd control in the city of al-Diwaniyah.

Not only did my squad continue working crowd control, this was also only the beginning of many incidents that involved distributing money to the Iraqi populace. After the initial payment at my bank, the Army soldiers came by on a regular basis. Before long, my squad eventually moved out of the bank and consolidated with the entire company outside the city in an old military camp. The 1st Marine Division had been camped there with a few other infantry battalions, but those units were soon gone, and it was just my company located at this particular spot. Despite being pulled from al-Diwaniyah, we were not relieved from conducting patrols in the city or rotating back into the bank to pull more security details. In between the rotations in and out of the bank and conducting patrols outside of the military compound, my squad was called from time to time to do quick reaction missions to quell crowds from forming in different areas on numerous occasions. This was hard work that put Marines face to face with the local people and often times, in attempts to keep the peace, ended with a lot of Marine aggression toward the locals.

About a month had passed since the first time the Army soldiers had come to my bank to pay civilians. It was about the first or second week of June 2003, and there had been several changes to the company. The main changes were that the Marines who were issued stop-loss orders had been sent home and replaced by new joins fresh from the School of Infantry.[6] The Marines showed up on white buses that had been driven across the desert, from Kuwait to al-Diwaniyah. Looking back, I cannot believe how naïve we were to have allowed an unescorted white bus to drive across Iraq. In any case, the Marines who replaced my squad members were well-trained, disciplined, and hungry for action. I told them that the squad had been doing mostly crowd control and that it was hard work. I believe crowd control is, in some ways, harder than pulling a trigger, so I tried to explain how they would sometimes have to be overly aggressive with the local people to help them understand what was expected from them. The new Marines had been there for only a week when we were called to conduct a mission.

On one afternoon, we were rushed into 7-ton trucks to go from our camp to a bank in the city. It was a different bank but only about three miles down from my old bank where my platoon's 2d Squad was located. If I remember correctly, 2d Squad was there as a part of a squad rotation that was circulating through the platoon. Apparently, something very similar to what happened to my squad was occurring at 2d Squad's location. The Army came in with money and was going to pay another group of people. As expected,

---

[6] *Stop-loss* refers to the president of the United States' authority to extend an enlistment or period of obligated service or suspend eligibility for retirement of a member of the military during time of war.

the Iraqis flooded the compound, practically overrunning the bank area. The 2d Squad called for more support and my 3d Squad and 1st Squad hurried to arrive and help them. The bank was a significant two-story building with a 10-foot-high courtyard wall surrounding it. The bank building was practically in the middle of the courtyard, and there was a lot of room to maneuver around the building. As we showed up at the bank's courtyard gate, I could see the chaos, and people inside the courtyard were crowding around the right side of the building. My platoon commander ordered us to dismount the trucks and sent us inside to corral the people into a manageable crowd.

As luck would have it, the commander then ordered my squad to occupy the most crowded area of the bank, which was on the right side. The reason for the excessive crowding on that side was because there were about five slits in the wall just big enough to exchange money through, and a crowd of 100–200 people had assembled there, screaming and yelling to get paid. He gave me orders to spread my squad out across the right side of the bank and to push the people away from the slits. I grabbed four Marines for this job while my platoon commander took my other men to plug up a hole somewhere else. The small detachment with me then pushed through the crowd and spread out in front of the slits with about five meters between each of us. I pushed through the crowd first and was on the far end of the five-member detachment making it very hard to relay commands.

Once we were in position, the situation became even worse. We were swarmed by Iraqis. It looked like one Marine surrounded by a sea of people. It was hot in the middle of the crowd and smelled horrible. I was pushed in every direction trying to hold my ground. I looked to my right and saw Lance Corporal Shane E. Kielion. He was one of the new joins, and I am not sure if he was even a private first class yet, but he looked at me and said, "Sergeant, what do we do?" Immediately, I did what I had done in every crowd control situation thus far. I swung my weapon around until I built a small perimeter around myself. As I did this, I looked over and saw that Kielion and the other Marines followed suit doing the same. This would only work for a short amount of time. I do not think this was my idea, but after what seemed like an eternity, we finally came up with a better plan. We completely pulled out our detachment of five men and stood at one end of the crowd. The plan was to start at the end of the crowd with two Marines positioned in a small phalanx to push the crowd away from the wall and the slits. Then, we would post a Marine at each slit to keep the crowd back once we gained that separation. We executed this maneuver, and it worked perfectly. Even though the crowd was still unruly, at least it gave us the space we needed to keep the people from crowding the slits and confusing the bank tellers.

Though my detachment had achieved our initial goal, the courtyard was still in great turmoil. There were people everywhere, and I remember being relieved at one point and walking around the bank building. I saw people urinating and defecating in one section of the compound. In one corner, I saw a woman crying and yelling at what seemed to be her father not to put a syringe in his arm. I am not sure what was in that syringe, but it was just one more situation that only added to the confusion. I even heard a story of a woman who was trying to get paid but could not because her husband's other wife had already received the money for his family. As I walked and worked in the compound that day, I was disgusted with what was going on. Why did these people form a disorderly crowd, and what was the point of all this? Eventually, the day passed, and we received more reinforcements, and another platoon showed up to help. The decision was finally made to form a phalanx consisting of the two onsite platoons and to push everyone out through a large blast hole in the courtyard wall. The platoons executed this maneuver, and the crowd was finally out of the courtyard. It was a tough day, and my squad along with the whole platoon remained in place to prepare for the next day.

As the units learned during the time at the first bank, we needed to have a plan for the next day. This time, we would be ready and have more control of the people who would be allowed to come into the courtyard. My squad was sent to the military compound to pick up five large spools of Iraqi concertina wire. These spools of wire were five to six feet long and three to four feet wide. We brought the spools back and organized them on four points to form a square with some separation between each large spool to allow people to walk through. This square was set in place outside the courtyard wall where the blast hole was located. The plan was to have the Iraqis line up between the spaces of the concertina wire, creating five paths that corresponded to the five slits in the bank wall. The paths represented different towns from which the locals would get paid. In addition, we would try to keep all the people separated in organized lines. Lastly, there would be a Marine in the front of every entry space in the concertina wire to control who came in and when. The plan was meant to bring order to chaos so that we did not have a repeat of the day before.

When the next morning arrived, the platoon set in place and organized the lines of Iraqis to send them into the courtyard windows so the Army staff could pay them. We then let the people in, and the Army began paying. Unfortunately, the lines that we thought we would have outside turned into one big crowd. It was crazy on the outside of our square perimeter, just like the day before, but at least, it was outside of the courtyard instead of inside of it. People were being crushed and pressed up against the spools of concertina wire, and it was about 10 degrees hotter in the crowd. Eventually, I made my way to the square to relieve some of my Marines. There were no long breaks for anyone. Everyone was actively engaged and employed. I remember manning one of the entry paths when one Iraqi grabbed my weapon; I utilized my martial art training. I circled out of the grab and buttstroked him in the solar plexus, and he fell like a sack of potatoes. Emotionally, I felt nothing and immediately placed one of my Marines to secure the path. I needed to get away for a minute from the constant stress of continuously being angry and aggressive.

I was in and out of the square all day, and each time I saw something new and shocking. One time, while in the center of the square supervising my Marines, a Marine from another squad jumped on top of the concertina spools with a stick and started to smack people in the heads to get them to back off of the wire so they would not crush other people. This captured my attention because, at that moment, I realized that we were here to help these people, but they were no longer people in our eyes. These people had stopped being mothers, fathers, and children but had become a faceless, cold mob to be subdued. In that moment, I felt I had lost a piece of myself, a piece that I would not recover until years later.

During one of my most significant visits to the square that day, a huge Iraqi was riling up the crowd and causing problems outside of the square. He stretched out his long arms practically crushing the people around him and causing many Iraqis to be pushed into the spools of sharp wire. My platoon commander was there and saw how agitated the man was making the crowd. So, the commander turned to me and said that he was going to open the path in front of him just long enough to get this big Iraqi man away from the defenseless people, so I could snatch him up and kick him out. I knew I was going to need some help. Luckily, two of my Marines were behind me. I looked at them and told them what the plan was and that I was going to need them to help me take this guy down. The platoon commander gave the order to open the floodgate, and Iraqis came rushing in. Finally, the big Iraqi made his way in, and I pushed his shoulders to the ground until he was on his hands and knees. I then grabbed him under his neck in a guillotine hold. Just then, the big, strong Iraqi man began to stand up, and I knew I was about to get some airtime, so I decided to drop all of my weight on him. He then collapsed back down, and somehow I turned him to his back. I dropped my knees on his stomach and heard the breath being knocked out of him. This was all done within sight and earshot of all the Iraqis in the crowd. As they saw me do this, the crowd became inflamed, and I knew I had to get this guy out of there. I quickly flipped him back over to his stomach, locked his hands together, picked him up off the ground, and walked him inside the courtyard.

My platoon commander and the interpreter then talked to the man and told him he would be paid that day. With that done, I went back out into the square and asked my two fellow Marines, who were supposed to help me, where the hell they were when I was fighting with this huge Iraqi. One Marine said, "It looked like you had the situation handled just fine." As the day came to a close, things calmed down, and we sent the crowd away to be paid another day. These crowd control situations were intense face-to-face confrontations with the people and were extremely difficult to deal with. I know these Marine units and tactics created a lot of enemies for future Coalition forces to deal with, but we did our best with the training, guidance, and knowledge we had at the time.

The combat stories mentioned above were written to the best of my knowledge. Many years have passed from when these experiences occurred. These situations were not conducted under enemy fire but were an integral part of the mission of the Marine Corps, and a lot can be learned. When reflecting on these events, three things come to mind: counterinsurgency, cultural awareness, and the transition from combat to peacekeeping operations. Over the years, I have reflected on those moments and wish I would have known more about these topics before going to Iraq.

Prior to 2004, little was discussed in my Marine company regarding counterinsurgency. I have only begun to get a true grasp about what is expected of a Marine in a counterinsurgency fight. I wish I had been exposed to counterinsurgency doctrine and directed to read books, such as those written by Mao Tse-tung or Ernesto Che Guevara, that discuss guerrilla warfare.[7] I also wish someone would have directed me to read about previous counterinsurgencies of other eras, such as Robert Taber's book *War of the Flea* (1965). I understand that it is hard to prepare for everything, but most of my training was strictly focused on patrols, fire maneuver, and fire and movement, and no time was spent on counterinsurgency. Everything in my training, up until that point, was centralized on killing the enemy and treating everyone with a heavy hand. One of the Marine Corps' standard operating procedures in training prior to Operation Iraqi Freedom was that, after conducting an ambush, we would sweep the kill zone and kill anyone who was still alive. A mentality like that is not very good for someone conducting a counterinsurgency operation. I still think many Marines are confused about what exactly is expected of them in a counterinsurgency fight. I saw this when I was teaching at the Staff Noncommissioned Officer Academy in 2009–13, and the war had been going on for almost 10 years. I believe that if I was at least exposed to counterinsurgency earlier, then I might have been more tuned in

---

[7] Mao, who was a Chinese Communist revolutionary, wrote *On Guerrilla Warfare* (1937). Guevara was an Argentine Marxist revolutionary and guerrilla leader, who wrote *Guerrilla Warfare* (1961).

to what the people needed and might have created fewer enemies for future forces. I am almost positive that many of my actions as well as my Marines' actions created more enemies.[8]

In addition to counterinsurgency training, I would have wanted more cultural awareness training. This also ties in with counterinsurgency. Often, I felt like my squad and I were lost and had no idea how we were supposed to engage the people in a "win the hearts and minds" fashion. Even if I did, I would not have known whom to communicate with in the populace. Throughout my deployment, I knew nothing of what a sheikh or an imam was and had no knowledge that Iraqi towns had city councils. I also did not know what part each, the sheikh and the council, played in the city. I did not learn any of that until I went back to Iraq on my second tour in 2007, which was three years later. I believe that if everyone had understood the culture and the roles these people played, things would have operated a lot smoother during my 2003 deployment. We could have already conducted leadership engagements and streamlined everything along the line of operation/line of effort (LOO/LOE).[9] Knowing the culture and roles of key individuals would have given my unit and its leadership some direction on the tactical level, while the operational and strategic planners were working hard to solidify a final LOO/LOE. At least, I would have been engaging the right people and known who to maintain relations with to keep the peace in my assigned areas.

---

[8] The problems noted by 1stSgt Galvan are the subject of many scholarly publications and are part of a larger historical trend. The Cold War ended an era (1945–89) in which the United States had one clear, identifiable enemy—Soviet Russia. After the 9/11 terrorist attacks, American citizens and the nation's leadership became aware of a new enemy that was in fact an unknowable conglomeration of enemies. Violent nonstate actors do not have the political and territorial constraints of nations, nor are they easily identifiable. What Galvan experienced on the ground is the same struggle that upper-level military leadership faced as the nation shifted from the Cold War to the Global War on Terrorism (GWOT). At both levels, the military had to identify the new enemies and, thus, also try to distinguish the new allies. Mao, Guevara, and Taber examined different eras when the line between the good guys and bad guys was indistinct and leaders had to be more creative about how to fight a nation's enemies without creating new ones.

[9] According to the U.S. Department of Defense, LOO is a line that defines the interior or exterior orientation of a force in relation to the enemy, and LOE uses the mission's purpose to focus efforts toward establishing operational and strategic conditions.

Lastly, in my opinion, transitioning from major combat operations to becoming a peacekeeping force was the most important and difficult objective to obtain. In my training, I only knew how to close with and destroy the enemy. I did not know how to switch that mentality off and go into a hearts-and-minds mind-set. I also did not understand what that meant. I had been told about the three-block war concept coined by General Charles C. Krulak, but I did not understand what it meant.[10] I only knew how to be aggressive and mean to everyone, no matter friend or foe. I wish I would have understood how to transition better from a combat force to a peacekeeping-type force. The three-block war concept, to my understanding at that time, was that I would be handing out food on one block, performing more kinetic operations on another block, and then engaging in full-on combat on the next block. I did not understand that my attitude also had to change from block to block; I should not have had the same attitude when I was in the combat block that I had when handing out food on a different block. In my case, during the time of these incidents, I had the combat mind-set the whole time. I can see why that is good for a Marine to have that ability, but once things turn to stability and counterinsurgency operations, that attitude can also cost a unit a lot in the short term and create more enemies to fight in the long term.

The experiences that I described were difficult. I remember my hands bleeding from holding my weapon at port arms for hours while I performed crowd control security.[11] I remember aggressively hitting people of all ages and sexes with my rifle to push them back and calm the rioting crowds. These experiences have been the hardest to deal with, even harder than when I had to pull the trigger against the enemy. I was inexperienced and had to make it up as I went along. I discussed this situation with my former company commander, and he agreed that squads like mine were not given

---

[10] Krulak coined the phrase three-block war in a January 1999 *Marine Corps Gazette* article, "The Strategic Corporal: Leadership in the Three Block War." Krulak referred to the mounting expectation of the Marine Corps to face conflicts in which Marines would confront the entire spectrum of tactical challenges in a short amount of time and within the space of three city blocks.

[11] Port arms describes a position in the manual of arms in which the rifle is held diagonally in front of the body with the muzzle pointing upward to the left.

a lot of direction. Not much was discussed about stability and counterinsurgency at the time of these operations; however, he was proud of what our company of Marines did in the streets and the results gained from the aggression we showed. My commander and I agreed that we did not know any better because all we knew was how to locate, close with, and destroy the enemy. We survived on improvisation and small-unit leadership, which sometimes proved to be great and sometimes proved to be dangerous. I can only hope that people who read this story understand the importance of training and educating military personnel on appropriate mission-oriented operations. A leader cannot plan for every contingency, but a Marine without direction can be extremely dangerous.

REFLECTION POINTS

- First Sergeant Galvan provides an honest portrayal of his experiences during a period of transition for the U.S. military. How have U.S. forces and other militaries been hampered by fighting wars based on the lessons learned from previous wars? Is there value in the lessons-learned approach? How can military leaders adjust training and education to overcome the unknowns of future wars?

- Using the experiences of the author, how is it possible for a Marine, soldier, or sailor to succeed and fail at the same time during a military operation?

- If you have served, how have you dealt with conflicting emotions or objectives during operations?

CHAPTER NINE

THE ADVISOR
COLONEL DANIEL L. YAROSLASKI

Colonel Daniel L. Yaroslaski is an amphibious assault vehicle (AAV) officer who has spent his entire Marine Corps career planning and conducting military operations. He has spent an extensive amount of time working with foreign military personnel from more than 20 countries. Beyond his unit deployments, Colonel Yaroslaski has had the opportunity to spend nearly three years of his career in formal education, including the U.S. Army's Armor Officer Advanced Course and the Marine Corps University's Command and Staff College, School of Advanced Warfighting, and Marine Corps War College. This time has allowed Colonel Yaroslaski to reflect on his experiences and make sense of complex issues, including leadership, ethics, and operational design. He served as a leader at various levels for the U.S. military and his combat experiences include tours in Iraq and Afghanistan. He coauthored two book chapters on the topics of ethics, spirituality, and sense making in complex situations. In this chapter, Colonel Yaroslaski discusses his experience as an advisor in Afghanistan and the ethical challenges he encountered.

From November 2008 to August 2009, as a lieutenant colonel, I had the pleasure and challenge of being an embedded training team (ETT) leader for an Afghan National Army (ANA) *kandak* (the term for a battalion). An ANA kandak is organized along similar lines with a Marine Corps or U.S. Army battalion with about the same number of personnel: three infantry companies, one weapons company, and a headquarters company. The standard tour for an advisor team was nine months as opposed to the normal seven months for most Marines; my assumption was that the longer period allowed for a more familiar working relationship with the ANA. Additionally, Marine Corps advisor teams in Regional Command-East (RC-East) were paired with ANA kandaks, but RC-East was manned with U.S. Army maneuver battalions (infantry, artillery, and aviation). This essentially placed the Marines Corps advisors in a completely new culture—a mix of Afghan and U.S. Army.

Every deployment is stressful not only for the servicemember but also on the servicemember's family, and this deployment was no different. Some of the unique stresses of this deployment in particular, though, stemmed from the fact that I was the commanding officer of Combat Assault Battalion, 3d Marine Division, in Okinawa, Japan, in addition to being the ETT leader. This meant that I was not the only one with commanding officer responsibilities, and my spouse was also now going to be highly involved with the deployment as an informal leader of Marine families. Normally, entire units deploy, but this time, things worked out a bit differently. In preparation for a deployment, military units are structured to allow for a large family support structure for the spouses and children during the deployment. Taking 21 members of the battalion with me meant that the family support structure was completely different and did not include as many spouses and children to form a large cohesive group. While the families of our small team bonded, my spouse still had a leadership role in the overall family support network while also working through the separation issues in our family. For the nine months that I was gone, she had to help the family readiness officer plan events and hold family support meetings for families who were "complete" while hers was not.[1] I know that added to her stress and the stress of our children, and I will be eternally grateful for the strength I felt from her as I served in Afghanistan.

This deployment was also different for me because I was not taking my entire unit; in fact, I was largely giving up command to my executive officer. This transition of authority did not necessarily translate to a transition of the responsibility I felt for the Marines and sailors of the Combat Assault Battalion. I had spent my career earning the opportunity to serve as a battalion commander, and the responsibility that comes with selection, while never a burden, is often times overwhelming.[2] I now had to come to grips with leaving the battalion behind and focusing on the mission I was assigned, which was to help make our kandak the finest fighting unit in the Afghan Army so that the United States and NATO could bring our forces home and Afghanistan could stand as an independent and viable country.

Perhaps the final stressor was the loss of a friend of the Marine Corps community in Okinawa suffered a few months before the deployment. A good friend of my family, Lieutenant Colonel Max A. Galeai, was killed in Iraq while participating in a town hall meeting.[3] A suicide bomber dressed as an Iraqi policeman infiltrated the meeting, moved close to Max and the official party, and detonated a suicide vest, killing Max and the town leadership. Max and his family had been neighbors in base housing on Okinawa; our spouses were good friends, our children were good friends, and my spouse taught Max's children piano. His death, particularly the manner in which he died, sent a shockwave through the ETT families. The ETT Marines knew that we were going to be in town hall meetings in Afghanistan, just like Max was in Iraq. We knew that if we were going to be successful in our mission, we would have to take the same risks he did, sitting at the head table without

---

[1] The FRO for a unit is responsible for the Unit, Personal, and Family Readiness Program (UPFRP), which includes disseminating command information, resources and referral information, deployment support, and volunteer management for Marines and their families.

[2] The term *selection* means that a board of senior Marine Corps leaders conducts a review of all candidates and selects the most highly qualified for command.

[3] Galeai was commander of the 2d Battalion, 3d Marines, out of MCB Hawaii when he was killed on 26 June 2008 in al-Anbar Province. Mary Vorsino, "Marine Lt. Col. Max A. Galeai," *Military Times,* Honor the Fallen database.

our helmets, weapons, and body armor, prepared to give our lives for the mission. Normally, those types of bold thoughts were simply concepts, but as we deployed to Afghanistan for our nine-month mission, those concepts became a harsh reality for the team, and more importantly, for the families we left behind in Okinawa.

THE PEOPLE, THE KANDAK, AND THE U.S. ARMY

The kandak initially operated out of three company-size FOBs in Kunar Province. The FOBs were within an hour-and-a-half drive from each other and were strategically positioned to allow Afghan and Coalition (in this specific case U.S. Army) forces to interdict insurgent activity and movement from Pakistan through five main valleys, across the Kunar River, and then onward into the other Afghan provinces. Each of the company FOBs was within six to eight kilometers of the Pakistani border and also sat at the foot of the Hindu Kush mountain range. Therefore, the Taliban could cross over the border using old goat paths and smuggling routes that enabled them to defeat some of the most sophisticated observation technology. Almost daily during the spring and summer and into the late fall, the Taliban crossed the border in either direction, and from the high ground, they fired 108mm rockets at Pakistani or Afghan/Coalition bases. Their movements were undetectable, and their ability to cross the border limited the ability of either military force to respond with gunfire. In other words, the insurgents had found a weakness in the political and military boundaries, and they exploited that seam with great skill.

The history of the Kunar River valley, both ancient and more recent, was a vivid part of the ETT members' daily lives. The local residents were fiercely proud about many things. In particular, they provided crops and hardwood to the rest of Afghanistan, and they had played a major role in defeating every invader from the British to the Russians. Their long history that included both intermittent periods of wealth and continuous conflict seemed to have developed a people with opportunistic loyalty and a desire to amass wealth to help carry them through the more common lean times. They had not developed a mercenary mentality; on the contrary, there were countless numbers of people in our area committed to building their villages and, by extension, Afghanistan into a resilient country with a sustainable future.

Figuring out how to help, who to support, and how to ensure money and humanitarian supplies went to those who really needed them was a source of ethical dilemmas. While in training, and even in daily planning, the missions we conducted and our interactions with the local populace seemed ethically straightforward, but actually doing what I thought was right was far more challenging. I had to shift from my Western way of interacting and attempt to learn to do business in an extremely foreign manner. I constantly balanced efficacy and ethics, and at times, no matter what choice I made I had to live with the uneasiness that came with the compromise of fundamental personal principles.

In contrast to the local village and family nature of the population, the ANA kandak, advised by our team of 21 Marines and U.S. Navy corpsmen, was a veritable pickup organization from every Afghan province and ethnicity. The kandak had been trained in Kabul by a previous team and deployed to the Kunar Province in early 2008. By November 2008, the members of the ANA kandak, and particularly its leaders, understood the local population, developed a local intelligence / information source network, and were well on the way to independent operations and a full readiness rating. As impressive as the kandak and its members were, there was a constant sense that they were barely able to operate, due mainly to logistical challenges. More than any other factor, the feast or famine mentality described above was alive and well in the ANA and presented the ETT with its greatest ethical challenges.

I was keyed into the importance of the cultural issues with the kandak because of my USMC training and experiences. Early on in my career, I learned a perspective that put me in the proper mind-set and framed my time with the Afghans and really every culture and foreign person I have met. I have been extremely fortunate to have received outstanding training and educational opportunities while serving as a Marine. The formal training, designed to develop my ability to effectively operate in different cultures, has served me well, but one event in my life made a major impact on my mind-set, ethics, and approach to dealing with others. While I was a second lieutenant at The Basic School, I shared a four-person room and experienced all the personality conflicts that go along with having four type A personalities crammed into a small living space. In particular, two of my roommates developed a rocky relationship because one (an American Marine) seemed to feel a sense of superiority over the other (a Mexican officer) and did nothing to disguise his feelings.[4] Our Mexican roommate was trying his best to keep up with the course, but command of the English language was clearly a problem for him. After a while, the Mexican officer jokingly announced he was going to stop reading his assignments and only read the dictionary since he spent so much time doing that already. The roommate with the superiority complex immediately missed the joke and began making condescending comments. Our Mexican Marine Corps roommate, without skipping a beat, then said something that I found to be very insightful and that would be helpful to me in the future: "Just because I speak with an accent doesn't mean I think with one." Those words gave me great pause and have been my guiding thought as I encounter new cultures and people.

Armed with an understanding of the Afghan people and their army, and with my predisposition for approaching new people as intelligent and worthy of engagement, I started my tour as an advisor. During my time, two particular incidents stood out as tests of my ability to maneuver the murky waters of ethical decision making. First, I had to sort through the internal divisions of a village and how to identify the leadership and conflicts within those groups based on a number of cultural variables. Second, I often found myself mediating between the Afghan and American commanders and had to learn how to balance each group's goals and concerns.

BARCOLAY AND KUSKALAY: A QUESTION OF ETHICS AND EMPOWERMENT

The road that ran from Jalalabad, the capital of the Nangarhar Province, Afghanistan, through Asadabad, the capital of Kunar Province, and into

---

[4] This was an officer from the Mexican Naval Infantry Force, Mexico's Marine Corps. TBS classes at Quantico often have international students from foreign militaries who attend alongside the students from the U.S. Marine Corps.

Nuristan Province is a strategic line of communication, and the paving of the road with funding from the U.S. government greatly improved the quality of life for the Afghan people. Once paved, the road had an immediate economic benefit. Small businesses developed along the roadside, and farmers were able to move crops to market in the provincial capitals, even in Kabul, in what could be considered record time for that part of Afghanistan. Unfortunately, when the road was built, money was not included for the quarter-mile stretch of road from the main highway to the bridge over the Kunar River. Coalition and Afghan forces had to pass over that quarter-mile stretch multiple times every day, directly through the thriving village that had made the key intersection into its commercial center. The increased traffic quickly gouged the dirt road surface and created large potholes where standing water formed, making the conditions even worse. To the credit of the villagers, the unit was never once attacked from any of the buildings that lined the road, despite the narrow passage and high volume of Coalition and Afghan Army vehicles. My fear, however, was that good grace would quickly be eroded if we overturned a vehicle, hit a building, or, worst of all, struck a child with a vehicle that was attempting to avoid a large pothole. So my personal mission became the paving of the road.

I worked with the regional USAID representative to secure a $10,000 grant that could be used to repair the road if, and only if, it was done through local labor and for the improvement of the community and, most importantly, administered by village leadership. To meet those conditions, I asked my ANA counterpart to request a meeting (known as a *shura*) with the village elders. The kandak commander and I had a superb relationship, and I found him to be an extremely thoughtful and well-respected leader. During the Soviet-Afghan War (1979–89), he had been one of Ahmad Shah Massoud's principle lieutenants and was trusted with the defense of Kabul against the subsequent Taliban invasion.[5] Once summoned by the ETT and ANA kandak in December 2008, the elders eagerly responded, and as the village leaders arrived, one elder in particular, "Dr. Ali" who was in charge of the local vil-

---

[5] Massoud was a military and political leader of the Northern Alliance in Afghanistan, which opposed the Taliban, when he was killed in a terrorist attack in the country on 9 September 2001.

lage's young men's association, struck me as the kind of person who could take charge and provide some much needed work and income for the same demographic the Taliban was recruiting from.

I believe over the years I have become a relatively good judge of character, and in the stress of the shura and the desire to make the project happen quickly, I too quickly became focused on working with Dr. Ali and the young men's association. Unfortunately, however, by empowering Dr. Ali over the other elders, I had forgotten my guiding principle—just because someone speaks with an accent, it does not mean he thinks with one—a mistake I was definitely going to pay for. In Dr. Ali, I found someone who was ready to do business, who seemed to be a logical choice, and who, I believed, represented the village. Yet, in hindsight, I realize there were signals coming from the other elders that perhaps I was moving too quickly and putting too much power in the hands of one person. After the second meeting with Dr. Ali, the ethical dilemma started building. I had ignored the other elders and Dr. Ali knew it. So, he behaved in the feast or famine manner I described earlier, asking for more money and attempting to reduce the scale and labor involved with the project. Dr. Ali could see that this project would be a boost to his power and influence, which he could use to increase his wealth and standing so that when the project ended, he could be in a better condition to use that wealth and standing to survive less prosperous times. For my part, I had committed my reputation to this project and had secured the first payment from the $10,000 grant for supplies, so there was no turning back.

Initially, I wanted the ANA kandak commander to be involved with the project so the ANA would get credit for the road as opposed to someone from the United States receiving the recognition. Unfortunately, the first report I got from Dr. Ali was that ANA members had asked him for half of the money since they brought him the business opportunity. Dr. Ali would not say who asked, but I quickly realized that as much as I wanted to bring credit to the ANA, the actions of its members were making it impossible. I brought it up to the commander, and he suggested that I exclude them from the project to prevent any further problems. Once the ANA was excluded though, I was on my own in dealing with Dr. Ali. After about a month of delay from Dr. Ali, I met him at his pharmacy and pointed out nothing had been done. He reminded me that things took time and that he actually had to use a significant percentage of the first installment of the grant to pay off other village elders before they would allow their sons to participate in the project. As angry as that made me, I realized that I was dealing with a different culture, a different mind-set, and most of all a different economic system. With Dr. Ali's assurance that he would now start the project, I reasoned that although I felt it was unethical to make payments like that, the overall benefit of the road would outweigh the corruption.

As I was dealing with Dr. Ali and his questionable ethics, the other members of the original group of elders I had assembled visited me. I learned that Dr. Ali did not exactly represent the entire village. Dr. Ali was from the commercial center business shura, yet there were two other shuras—one from the upper village of Barcolay, and one from the lower village of Kuskalay. Members of those shuras also told me that Dr. Ali was excluding them from the project and that they did not want the road project to continue unless there were going to be more projects for them. Despite what seemed like blackmail, I was able to convince them that the USAID grant program funding the road project could fund lots of other village improvements too; they just needed to bring their projects to me, and I would work with them. One elder in particular seemed extremely hostile, and my instinct told me that he was a member of the Taliban. Unfortunately, he was also the elder who the others chose to be in charge of the next project. I had a lot of background work to do before I felt I could really trust him to receive U.S. financial support. By the end of the meeting, the elders were satisfied, but I was stuck in the position of potentially working with the enemy, which I was definitely not prepared for or willing to do.

Shortly after the meeting with the other elders, I received a call from Dr. Ali who said that the road had been graded, and the much-needed drainage had been dug. Yet, he had run out of money and was unable to pave the road. I made a call to a local road building company, and they confirmed that even they would not have been able to do the entire project for $10,000. That information helped relieve my conscience, and in fact, I even began a second project. The problems were not quite over yet, though. The ANA had again threatened Dr. Ali, this time for more money than the project was actually worth. As much as I wanted to confront the commander with this news, I realized it would only bring bigger problems. Thus, I decided to tell the kandak commander that the project was over and that the grant money given to Dr. Ali had been spent. I believe the commander understood what I was saying, but I certainly felt like I had taken the easy way out instead of righting the ethical wrong of the situation. Shortly after the conversation, Dr. Ali confirmed the threats had stopped.

To finish the road, I then had to have additional meetings with the village elder who I suspected was with the Taliban. I had to figure out who he was and what he really wanted. Thanks to the hard work of my interpreter, I learned that he was actually an honest individual, but just to make sure, I allowed him to administer the next $10,000 while I handled the actual payment distribution to the road construction company. Fortunately, the construction company agreed to hire local workers, which met the USAID requirements for the grant.

The overall road improvement and paving project took six months, cost $20,000, and required an enormous amount of time and energy. I learned some valuable lessons though. First and foremost, I needed to listen carefully and adapt to the local process. It may have seemed unethical, but in the context of the village elder system, it was effective and ensured peace. I also learned that I had to live with the ANA and its use of power for profit, which either meant I could be complicit with the ANA members or work around them, but what I could not do—no matter how much I wanted to—was change them. Finally, I relearned that by not "thinking in English" and not assuming my assessments of who was intelligent and effective were naturally correct, I would have more chances to succeed.

DISPROPORTIONATE RISK—AN ETHICAL CHALLENGE

While my team and I were working through the ethical challenges presented by the people of Afghanistan and the relationship with our ANA counterparts, we also had to deal with the challenges presented by the advisor mission and the Taliban. During the training and preparation for our mission, I often told the team that our job was not to rush in and lead the fight. Our job was to step back, assess, and help bring order during a fight and to figure out what support we could bring to bear from the Coalition forces. As members of the U.S. military who are subject to overall International Security Assistance Force (ISAF) rules, moving with and supporting the ANA when it needed us was sometimes challenging.[6] Rules like minimum numbers of vehicles in a convoy and minimum numbers of ISAF personnel in a particular location for any extended period of time were nearly impossible to comply with. Our team was only 21 members strong. Even with six extra personnel from National Guard units, in order to cover four separate FOBs, the team was spread extremely thin and had to make tough choices. Just accepting risk in the name of mission accomplishment would have been the easy solution, but it was not always the right solution.

The value of an ETT was the access to additional support it could bring to bear from the Coalition forces (air support, artillery support, and medical support were the most common requests). The burden of an Embedded Training Team on the ANA was an extremely heavy vehicle that the ETTs operated that, if disabled by enemy action or mechanical failure, was beyond the capacity of the ANA to recover. Additionally, the ISAF vehicles carried a bigger bull's-eye than an ANA Ford Ranger pickup truck. Explaining why I could not send ETTs on a mission was difficult, considering the chances the ANA members were prepared to take, with or without the team's assistance. Often times, it seemed like I was making excuses as to why the lives of my ETTs were more valuable than the soldiers of the kandak, and no matter how many good and operationally sound reasons I may have come up with, it just felt like a betrayal of what was right.

---

[6] ISAF was a NATO-led security mission in Afghanistan established by the United Nations Security Council in December 2001.

This risk calculation was similar to the law of armed conflict principle of proportionality in war.[7] Just as a military action that causes collateral damage must be proportional to the military value gained, the amount of risk an ETT took on a daily basis had to be calculated against the amount of risk the ANA would incur if the team was not there. Also, if the ANA stopped seeing the ETTs as valuable for anything other than a source of funding and fuel, our ability to influence real change would rapidly evaporate. Interestingly, a number of the older ANA officers had served on different sides of the conflict during the Soviet occupation. The ones who had been *mujahideen* were much more willing to risk their lives and the lives of their soldiers and had an expectation that the ETTs would do the same. The former Soviet-trained officers were more willing to exercise tactical patience but were also more than happy to let U.S. forces take the lead.

This tension associated with proportionality of risk also extended to the types of operations I recommended the kandak take part in. The U.S. Army forces in the Kunar Province were extremely well equipped, from the most modern mine-resistant ambush protected (MRAP) vehicles to thermal night vision goggles to highly effective cold weather gear and clothing. The ANA, on the other hand, mostly had one set of uniforms, a cotton field jacket, old sleeping bags or blankets, and more importantly no night vision devices. In fact, this lack of night vision equipment meant the ANA was reluctant to operate at night away from buildings or without the headlights of their Ranger pickups. The mismatch of capability led to numerous confrontations between desire for action by the Coalition force and the resistance of the ANA to embark on operations that the ANA soldiers were clearly not equipped for.

One of the main skills I quickly realized a good ETT required was that of interpreter—not of language, but of intent. Perhaps the skill could also be called mediation, and I always felt like a man trapped in the middle. The

---

[7] On page 61 of the 2015 U.S. Department of Defense *Law of War Manual* the rule of proportionality during war is said to oblige a force to refrain from attacking when the "expected harm incidental to the attack would be excessive in relation to the military advantage anticipated to be gained."

conflict generally occurred along a few familiar patterns. First, the U.S. forces would receive an intelligence report, often at night and always on very short notice. The intelligence was good and yet not good enough to send a patrol, so I would get the call to send the ANA. The first few times this happened, I immediately went to the kandak commander and told him, "We have to take action. The Coalition force needs us." Often times on particularly dark and cold nights, he asked if Coalition forces would also be going. After awhile, I realized his question was code for "there is no way we can do this mission with the soldiers and equipment we have and get the results the Coalition wants from us." As with the villagers in my earlier example, I did not hear the code right away, and I pushed to execute the Coalition requested operation. As you might expect, and as the kandak commander knew, the operation turned into a nighttime drive to nowhere, ending without the result the Coalition wanted. It was really an expression of exercising ego more than anything else. The ANA commander knew if he said no to the mission, he would look weak to the Coalition and reports would get back to ANA leadership that his kandak was of low quality.

With a few of these experiences under my belt, I finally learned to hear the code. I stopped the request from going to the ANA and then, on the next day, explained to the Coalition leaders why I did it. In actuality, I was taking the ANA code and interpreting it into U.S. Army code—language that a U.S. Army battalion commander could understand. The requests never really stopped, and I learned how to push the ANA commander to say "yes" to the ones that were simply impossible to refuse without harming the kandak's reputation.

The second type of conflict arose, oddly enough, when the ANA wanted to conduct operations that the Coalition did not want to participate in. The ANA commander had a highly developed information source network in the Kunar Province. The problem for the Coalition, though, was that the sources were not vetted and, therefore, could not be considered reliable. This lack of reliability, particularly when a source called in the middle of the night in a remote location, caused the U.S. Army battalion to have to say "no." This odd reversal of requests meant that a high number of ANA source reports were never acted on. Again, my job was to interpret the code and translate for both sides in order to find a common ground. Generally, the kandak commander called my interpreter, or *tajiman*, who woke me up with the news of a source report. I went to the Coalition Command Operations Center and relayed the report to the battle captain or intelligence watch officer. Those individuals would check their validated reports and, when they did not find valid reports, would state they could not act on the ANA report. I then had to return to the kandak commander and tell him the Coalition could not act. Without him saying anything, I knew the kandak alone would not be able to act. By the end of my nine months, I had become adept at explaining both sides and, more importantly, reporting through the advisor chain that if properly equipped, the kandak showed every sign of being an outstanding organization.

The code often times required extra work on my part. This was particularly true when the camp came under fire from rockets and the ANA sources started giving rocket launch locations. Without two forms of positive identification (PID) from valid sources, counterfires could not be requested. The brigade commander had given the kandak commander explicit instructions that, if he knew where an attack was coming from via information from a reliable source, the kandak was to use its D-30 122mm howitzer artillery to return fire and defeat the attack. The first few times the kandak wanted to use the D-30s, the kandak commander seemed to be asking my permission; in reality though, he wanted me to clear the fires with the Coalition to ensure there was no fratricide. Again, before I learned the code on both sides, I asked the Coalition for clearance, but without PID, they could only say "no." Once I figured out what was going on, we conducted gun drills five to six times per week, demonstrating the crew could hit a target consistently and with a high degree of accuracy. I also made sure the Coalition force leadership from the local battalion to the brigade had observed the crews and knew they were proficient. Finally, I asked the ANA brigade commander to talk to the Coalition commander and let him know that the crews were trained, that the intelligence was reliable, and that the D-30s would be used only against the enemy firing points that were known not to have civilian structures (such as the open area on the Hindu Kush that was perfect for launching rockets). More important, when the time came to fire, the ANA would ask if the area was clear but would not ask for permission to fire.

Once both sides learned each other's code, and I learned how to listen and translate, the ANA and Coalition were able to execute in completely new ways.

We were not always successful, nor were we always operating smoothly, but we were always working together.

From my time in Afghanistan as an advisor, I learned some new lessons and relearned some old ones. An advisor must balance proportionality between mission accomplishment and risk and be ready to take charge, but must first step back and assess the situation. Most important, though, whether dealing with the local population, the ANA, or the Coalition, the words of my TBS roommate have served me well in my role as an advisor. Just because someone speaks with a Dari, Pashtun, or even American accent, it does not mean he thinks with one.

REFLECTION POINTS

- How does the cultural bias among groups involved in war affect the interactions of military servicemembers and civilians? How can those perspectives be exploited?

- How can ethics be culturally and situationally based? Is compromising a leader's ethics ever acceptable? In other words, can the ends justify the means, or are there other ways to obtain results?

- What benefits and challenges come with combining military forces nationally and internationally? Do the advantages of working with other agencies and military forces outweigh the problems caused by doing so?

CHAPTER TEN

A LOGISTICIAN'S PREPARATION FOR COMBAT
LIEUTENANT COLONEL CLIFTON B. CARPENTER (RET)

Lieutenant Colonel Clifton B. Carpenter (Ret) served as a Marine Corps logistics officer from 1993 to 2013. His career included execution of the Lebanon noncombatant evacuation operation (NEO) in 2006 and a 10-month deployment to Helmand Province in southern Afghanistan with the 24th MEU, which commenced the Marine Corps' transition from Iraq to Afghanistan in 2008. Lieutenant Colonel Carpenter also served as the primary operations briefer to the Commandant of the Marine Corps and commanded the Logistics Operations School at Camp Gilbert H. Johnson, North Carolina, from 2010 to 2012. Lieutenant Colonel Carpenter details his deployment experiences as a major with the 24th MEU.

Of our lost comrades in arms, President Ronald W. Reagan said, "They gave up two lives—the one they were living and the one they would have lived. When they died, they gave up their chance to be husbands and fathers and grandfathers."[1] We honor our lost comrades in many ways. One of those

---

[1] Ronald Reagan, "Remarks by President Ronald Reagan at Veterans Day National Ceremony" (speech, Arlington National Cemetery, Arlington, VA, 11 November 1985).

ways includes preparing the next generation of warriors by writing the lessons learned from previous generations. As I read the stories in this book and in other recollections of combat, I had a sense of awe for the human and American endurance seen in them. I am humbled by recounts of bravery and sacrifice for brothers and sisters in the heat of combat and by the scars of the mind and body our veterans endure.

My own story includes no Purple Heart at the end. I have no tale of personal bravery or physical injury. I was rarely at risk of casualty during my 20 years of service, including two combat deployments. I was a Marine logistics officer; this task, when done correctly, garners little appreciation and, when done poorly, becomes the scapegoat for mission failure. As Alexander the Great is attributed for saying, "My logisticians are a humorless lot . . . they know if my campaign fails, they are the first ones I will slay."

PERSONAL BACKGROUND

Unlike many young officers who took their oaths of office after 9/11, I was a major with 14 years of service when my unit was directed to Afghanistan. My story began in 1992 when I attended OCS the summer before my senior year at the U.S. Naval Academy. Those six short weeks sold me on my Service selection choice, and I found myself wearing the Eagle, Globe, and Anchor on graduation day.[2] To me, the Marine Corps was a team. Being a member of the Corps felt comfortable. As an athlete, I had learned the traits that the Marine Corps taught as sacrosanct ideals, including training for endurance to achieve a common purpose and upholding the virtues of honor, courage, and commitment. The logistician part was ancillary. I had wanted to be a grunt, like many young lieutenants. When my staff platoon commander read my future—logistics vocation—I was not sure what the word logistics meant. My new strategy for 1993 became "five and dive," or in other words, I had to serve the five years that were required for Naval Academy graduates, and then, I planned to separate from the military for greener pastures.

Fifteen years later, I was looking back at my tours, which began in 1994 at Camp Lejeune. I had been sent to Pohang, South Korea, for a year in 1997

---

[2] Eagle, Globe, and Anchor is the official Marine Corps emblem and insignia.

and then to the Mountain Warfare Training Center in Bridgeport. I was at the Amphibious Warfare School (AWS) at MCB Quantico in 2002 and spent two years with the Marine Corps Systems Command.[3] In 2005, I was at the Marine Corps Command and Staff College before my assignment to the 24th MEU.[4] In 2006, the 24th MEU conducted an amphibious deployment and executed the Lebanon NEO.

So, there I was, having not five and dived, 15 years later. In January 2008, I found myself as a major on my second deployment as the logistics officer for the 24th MEU. I was rightly expected to know how to move the unit around the world and how to sustain it when it got there. Further, through two deployments, as with any MEU, I was surrounded by mid-grade, competent captains and staff NCOs, who knew their jobs. A MEU is a chaotic but elegant instrument made up of many parts that work as one. My story does not reflect on preparing for the horrors of war; other veterans are much more qualified to comment on that topic. Nor is this an after action report (AAR) for logistics support in southern Afghanistan; again, there are many versions of that. This book is about individual preparation for combat, not collective technical skills. With that perspective in mind, my goal in this chapter is to reinforce how three aspects of combat preparation help to prime Marines for war. From my view as a Marine leader and a logistician, I highlight three topics—combat laboratories, families during deployments, and a leader's reliance on the team versus reliance on self in a combat organization—supported by my personal experiences and presented as topics that are typically less explored than others in collections of combat reflections.

COMBAT LABORATORIES

The Marine Corps' most requested contribution to various combatant commander's force requirements is the MEU. Prior to 9/11, six MEUs (three on the East Coast and three on the West Coast) trained for six months,

---

[3] In 2002, AWS was renamed the Expeditionary Warfare School (EWS).

[4] A MEU combines a headquarters, commanded by a Marine colonel, with a reinforced infantry battalion, a composite aviation squadron consisting of helicopters and jets, and a logistics battalion. The unit traditionally deploys onboard U.S. Navy ships.

deployed for six months, and rested for six months; the Okinawa MEU has a little different schedule but a no less challenging mission. For the first 10 years of my career, the MEU was the organization everyone wanted to join because the unit's training was well funded and it always deployed. Other units would count their most interesting adventure as a trip to the combined arms exercise (CAX) in Twentynine Palms. The attacks on 9/11 changed that attitude because all combat formations had great training opportunities (e.g., Exercise Mojave Viper) and real-world combat deployments. As Marines look forward post-OIF and -OEF, I submit that a return to a MEU focus supports the Corps' roots as an amphibious force, it returns to the mind-set of self-reliance in preparation for combat (as opposed to strategic deployment by the U.S. Air Force or contract aircraft or ships), and most importantly, the MEUs are the best peacetime training laboratory the Marine Corps has and must remain an integral part of the Corps to properly prepare Marines for deployment.

The training a logistician receives at a MEU is irreplaceable because there is no way to simulate the cycle of deployment, sustainment, and redeployment. Logistics supports effective movement and support of combat formations, and like being able to do 20 pull-ups, success comes through doing it. Additionally, as a Marine air-ground task force (MAGTF) officer, I believe the integration of warfighting functions that occur in a MEU deployment cannot be replicated; this is why the MEU program remains essential to the Corps' MAGTF identity and training.

THE OTHER MAGTF ELEMENT: FAMILIES DURING DEPLOYMENTS

Combat tests the very limits of individual physical and mental preparation. The Corps has always been good at the physical preparation of Marines. From 2001 to present, I believe we have improved significantly in the emotional preparation of Marines as well. No less critical is the preparation of Marine families. This process is less trainable and less measurable, and deployment complicates matters for families, especially since every family has different needs. I will talk about mental and physical preparations my family endured, including the lessons my wife, also a Marine, and I learned while preparing for four combat deployments (two for me and two for her) and how we tried to incorporate those lessons into preparing our Marines' families for combat as well.

A LEADER'S RELIANCE ON TEAM VERSUS RELIANCE ON SELF

Finally, combat preparation requires what I call the "individual effort break point" for leaders. I propose that there exists, for every officer and staff NCO, an intersection where the efforts of the individual and the team meet to achieve the best outcome. In contrast, there is a point where individual hard work can no longer replace the effort of a functioning team, which is the break point. The sooner Marines and Marine leaders learn to trust their teammates, the more effective they will become when the true test comes in combat. So, training in the MEUs, preparing families, and acknowledging individual effort break points, as well as how those things relate to individual preparation for combat, are the topics in explaining what it was like to go to war.

GOING TO AFGHANISTAN

It was 5 January 2008 in Camp Lejeune. The office was empty. Despite being only 20 days away from embarkation onboard amphibious ships for a seven-month float, or tour of duty, everyone was enjoying the last of a New Year's holiday, and the 24th MEU was ready to go. The unit's predeployment training plan was complete, except for one certification event, and the unit's cargo was already embarked. This allowed the MEU commanding officer to send everyone home for the Christmas and New Year's holidays except a few unlucky folks who were watching cargo in the holds of three ships. The MEU commander and I had grown on each other through the course of the predeployment work-up. He was a very intelligent officer with years of combat experience, the up-close and personal kind of experience that drives people to become very confident, but he was mostly quiet. What he did not have was a great deal of experience with the sequential machinations required to move and sustain large units, nor did he have a great deal of patience for listening to staff officers talk about logistics. He liked to talk about operations with his operations officer and handguns with his artillery officer. Logistics nauseated him. In short, he was stuck with me. To his credit, he left me alone to do my job and he let me know when I did not do it well enough, without trying to tell me how to do it.

I was the MEU's logistics officer (S-4) and was in the office on 5 January trying to move the piles of paper that had accumulated over the six-month predeployment work-up, when a previous commander called from the Pentagon. "Afghanistan," I remember saying. "When?" Thus, the hardest 90 days of my 20-year career began. From the time I received that phone call until the day a reinforced Marine infantry battalion advanced on the enemy in a hot, uncomfortable spot in southern Afghanistan, these events tested all I had ever achieved, professionally and personally. As a point of pride, ultimately my commander and my unit were able to execute the mission.

However, we collectively, and I personally, made plenty of mistakes. There were miscalculations that had to be rethought and fixed midstream. There were hundreds of questions with no answers, and like any good planner, I simply had to make assumptions and execute even though a lot of those assumptions were incorrect. I firmly believe that the flexibility shown by the MEU on this occasion, and by many other MEUs that have been asked to accomplish similar feats, rested on a cadre of Marines who knew how to move the MAGTF animal. It is one thing to say "send in a MEU" on a PowerPoint slide at the Pentagon. It is another thing entirely to watch all its pieces and parts slinging around the globe. Our ability to do this was possible because of the institutional knowledge retained by the Marines who had done it before.

TRAINING MEETS REALITY

Studying the deployment of forces bore little resemblance to deploying them to Afghanistan in 2008. I clearly remember a classroom session in logistics school that listed a legion of requirements to get transportation resources to move a force—requirements like accurate equipment inventories and passenger manifests 60 days prior to departure. In January 2008, the complete lack of an operational mission in Afghanistan compounded with a "hard date" for arrival and a unit configured for amphibious deployment instead of strategic airlift made for a dizzying nightmare. I will highlight two particular challenges: equipment transportation and money. The skills to manage these two areas are retained in the Corps by Marines with MEU experience.

EQUIPMENT TRANSPORTATION

A MEU's equipment configured for amphibious deployment does not look like a MEU's equipment for Iraq or Afghanistan missions. An amphibious MEU is constrained by the space available on three ships and prepared for a wide range of missions. A MEU going to Iraq or Afghanistan will conduct counterinsurgency operations and try not to get blown up by IEDs. The latter MEU is also much heavier with twice as many vehicles. And, by 2008, the MRAP vehicles were beginning to come off the assembly line, and none of these types of vehicles could fit on a ship. Furthermore, on an amphibious deployment, the ship itself houses all the intermediate maintenance capabilities for the aircraft, including tons of heavy stuff, electronic stuff, and stuff that needs power, water, hydraulic pressure, and more. Once apart from the ship, a MEU's aviation contingent must replace that capability. The unit did not possess any of these items. In addition, we had to provide the strategic transporters that were trying to meet the hard date for the unit's arrival, with the exact dimensional data for all the equipment to be sent to a still unknown location.

Were it not for the previous 24 months of history in my unit, I think success in this misadventure would have been unlikely. In the 24 months leading up to this scenario, my unit had deployed previously on amphibious ships. It had used rail transportation during training and strategic air in support of Hurricane Katrina humanitarian relief efforts. The unit had trucked itself, driven itself, and moved by contracted shipping. Each and every one of these methods was used previously, and each method had unique challenges. As I advocate for a return to a MEU-focused Marine Corps, I submit that another type of unit would not have succeeded in executing an unplanned, impromptu deployment for combat operations in a landlocked country. My unit's familiarity with all the transportation methods required to ultimately get us to Afghanistan came from having used them training and having the unique cultural flexibility that a MEU retains because it is a unit designed to be ready for an array of contingencies and is not focused on a single mission.

MONEY

How do young officers learn to spend the government's money? It is an intimidating topic because every young officer knows a horror story about a peer who was reprimanded, fired, or relieved for some fiscal blunder. Budgeting and spending money is another area where the staff and subordinate leaders learn by actually doing rather than from their training. A MEU has real money to spend similar to a division, wing, or group. Learning to spend money in the right way is no different than learning to fire a weapon or command an infantry company or fire a battalion's worth of artillery. To train for the technical skills normally associated with Marine units (e.g., shooting weapons, talking on radios, etc.), we learn incrementally. In a training setting like a rifle range, we control the learning environment so that mistakes are not dangerous, and we build experience. Unfortunately, there is no similar, controlled situation for learning to spend money. Often, officers are 20-year veterans and battalion commanders before they ever have any budget authority. Experience in the MEU is the closest thing young officers have for a money-spending laboratory. Each of the three components of a MEU—the infantry battalion, the aviation squadron, and the logistics battalion—have opportunities to spend money. Buying tools, fixing vehicles, and arranging for food contracts and garbage pick-up are all experiences young officers have in the MEU, which in another unit (a division, wing, or group) would be handled by a more senior officer.

My own experience with money as a logistics officer in the MEU prepared me for a very difficult experience in Afghanistan, one that tied the operational employment of the MEU directly to money. We were scheduled to enter Helmand Province in early 2008 before any other conventional U.S.

forces were operating there. Further, the pace of operational level change (the shift of focus from Iraq to Afghanistan) clouded the mission for which the unit would be employed when it arrived. Bottom line: we knew we were going somewhere in southern Afghanistan, but the theater commander was not yet ready to commit us to a particular location. For a logistician, this nightmare meant building infrastructure for the arrival of 3,000 Marines on a very short timeline without actually knowing where to build it. I distinctly remember a class discussion on how commanders validate requirements based on specific requests for forces that are justified by a combat mission. In January 2008, the unit's mission was shown as a giant pitchfork on a PowerPoint slide positioned over an area the size of Georgia. In other words, we could be expected to be employed in finding the enemy—for which little information existed concerning an exact location or activity—in a huge area, in a featureless desert of fine sand, and in well-over 100 degree temperatures: a logistician's nightmare (as well as an infantryman's or an aviator's). From a logistics perspective, in lieu of anything more definitive, the object became to provide options to the commander in hopes that clarity would come later.

For a logistician, flexibility and making options for the commander means buying your way out. My unit was still in the United States preparing to deploy, so my commander sent me with a group of planners to Afghanistan to make preparations for our arrival. I needed contracts for erecting buildings and engineering work so that aviation units could park and operate aircraft, and I needed contracts for services to feed and clean up after 3,000 Marines, fuel, air conditioning, hospital services, etc. The key question was where. Nobody could tell us whether the unit would be near Kandahar, Afghanistan, in flat sand, in the mountains of Uruzgan Province, or as far west as Herat all seemed to be distinct possibilities. With that perspective, I had to answer simple questions with nonsensical responses. For example, a simple bean counter question inspired the following frustrating conversation:

"Will the unit operate from location X?"

"Yes."

"Will the unit operate from location Y?"

"Also, yes."

"How will the unit (including aircraft and fuel and ammunition stores) be distributed among locations X, Y, and Z?"

"The entire unit including aircraft will operate from locations X, Y, and Z all the time."

"But, that's impossible."

"Again, yes."

Without operational guidance on where the commander in Afghanistan would employ the unit, there was no choice but to plan (and pay for) employment in multiple locations. Outfitting 3,000 Marines and their equipment to operate in one desolate place is an exorbitant proposition. Buying options for the commander is a function of understanding contract structure, the scope of the operation, and other options, as well as not being afraid of someone asking how much it will cost. In my unit, with the support of the commander and with two-and-a-half years of experience spending a lot of government money as part of a MEU, my team and I were at the confluence of these circumstances. Like the equipment transportation example discussed above, if we had not had the MEU fiscal experience to both understand contract specifics and gain the confidence to answer for the commander 100 unique and expensive questions, I believe the Afghanistan deployment would have been wrought with additional problems. In peacetime, the MEUs provide the only practical application of the wisdom contained in many books of doctrine.[5] The MEU is the cheapest MAGTF school the Corps runs with the highest output of confident officers and staff NCOs, and it is a capability liked and understood by the joint community.

Technical skills, such as transportation and finance, have to be practiced just like those more commonly associated with combat, such as shooting weapons and talking on radios. Flying an entire unit to Twentynine Palms to operate equipment that is already there has its efficiencies. It does not train unit mobility, and when the perfect storm hits, the Corps needs units that know how to prepare for movement and execute it instead of relying on other means for getting to the fight. One way to do this is to deploy units to

---

[5] Marine Corps doctrinal publications contain the Corps' fundamental and enduring beliefs on warfighting and the guiding doctrine for the conduct of major warfighting activities.

their training opportunities. By this, I mean that there are multiple ways to take a unit from North Carolina to Virginia for training.

The easiest is to have contractors put all the equipment on trucks and the Marines on buses. Everything gets to Virginia in a couple of days with little chance of equipment damage. The other way is to prepare the equipment and have the Marines drive from North Carolina to Virginia in a convoy. This has all the inherent risks of drivers getting lost, having accidents with civilian vehicles, losing communication with a convoy, being overweight for particular bridges along the highway, and multiple other potential problems. I submit that these are the same problems my unit faced in traveling from Kandahar to Helmand Province in Afghanistan, and I also submit that my unit was prepared for this trip because my commander was willing to consider administrative movement to training sites as training. Likewise, with money, if Marines want a Corps that can do impossible things quickly, we need Marines who can efficiently buy options. In other words, mid-grade Marine logisticians in conjunction with mid-grade combat arms officers and aviators must have the confidence to spend a scary amount of money legally to maximize and retain options for the commander; this skill is learned, no differently than firing a weapon. Young officers need more rigorous exposure to contracting education and contracting options and less crucifixion when they make honest, unintentional mistakes (excluding gross negligence), because experience is the only true teacher.

The opportunity to practice these and other critical MAGTF logistics skills, at any rank from lieutenant to colonel, are in the MEUs. I recommend that all young logistics officers, who are preparing for combat, take advantage of opportunities to deploy with MEUs as many times as possible in as many different billets as they can. Any other job they will have in their careers will be less challenging as a result.

PREPARING FAMILIES

I move away now from the means by which we hope to train the next generation of individual Marines and the unit's they belong to for combat. My second topic is about Marine families. Advice for preparing physically and mentally for combat can be found in publications on combat histories. However, to date, the very difficult but critical process of preparing families has received less attention.

Regardless of MOS, actual combat tests servicemembers' physical and mental limits. This is a truism available in every published narrative regarding combat. I share nothing new, and I share nothing contradictory to those narratives. The only difference between this story and the others is that this one is mine. Preparation falls in three categories in my view: physical, professional, and emotional. For those who joined the Marines after 2001 but before 2013, I likely have nothing new to tell you about physical challenges. For anyone who joined after 2013 and wonder what combat will be like, believe all you have heard, all that you have been taught, and all that you can imagine about discomfort, sleep deprivation, and exposure to the elements. Wearing all your gear in brutal heat (or cold) and sleeping little will drain the energy out of the strongest and most committed Marines. The effects of discomfort, fatigue, and sleep loss can be offset by good physical fitness. Likewise, regarding professional preparation for combat, a Marine's knowledge about his or her job will be stretched to new levels when the unit is actually leaving for a combat zone. These conditions are difficult enough without considering the emotional energy needed to deal with family concerns.

Emotional preparation for combat, unlike physical and professional preparation, is more difficult to quantify. Success in emotional preparation is likewise difficult to measure and impossible to know when it can be categorized as completed. However, lack of adequate attention in this area manifests quickly in combat conditions. Deployed Marines in emotional stress due to conditions at home are a wreck in battle. A sick child, a depressed or disloyal spouse, or financial distress will take a Marine out of the fight as fast as an enemy bullet. Leaders need to consider the situations that they leave at home when they deploy and the situations their Marines leave at home as equally important parts of combat preparation.

In 2008, when I left for Afghanistan, I was a mid-grade major married to a mid-grade major intelligence officer. My wife and I have four children;

at the time, two were very young. My wife had access to my unit's classified website, so she knew essentially what was happening while I was deployed and when and where bad things happened in real time. She knew that things on a day-to-day basis were not fine. I knew that the home front was also a long, long way from being fine. But, our Sunday afternoon conversations often did not go too deep and typically went something like this:

Wife: "How's everything over there?"

Me: "Fine . . . Yeah, great—we're doing great. How's everything at home? How are the girls' and the boys' grades? Okay?"

Wife: "Fine. We are all doing great. We miss you. Talk next week, okay?"

She was living the life of a Marine officer while getting two children to daycare and supporting the extracurricular grind of two high-school-age boys. She had more than just the day-to-day struggles of single-parent life; one of our daughters was in and out of the hospital four times. We had our house on the market for four months with no offers and a different rental house without tenants for three months. She executed permanent change of station (PCS) orders from Camp Lejeune to Washington, DC, with the whole family lash-up while I remained forwardly deployed.

Every family, on some level, has similar or more challenging circumstances. My wife and I were able to cope for several reasons. As two majors, we were never in real financial trouble. We had been in a loving and mutually supporting relationship for eight years, and my parents lived fairly close by to help when things got really out of control. Without any one of those support pillars (money, relationship, helpful family), we would have had a much less durable emotional situation for getting through the deployment. Take away all three pillars, and we would have had the family instability and emotional roller coaster that makes Marines of any rank susceptible to combat stress. In my experience, I saw many Marines lacking one or all three pillars. That tenuous family situations are extremely common in Marine families should not be surprising. The majority are newly married parents with combined incomes in the lowest third range for Americans as a whole.

How then does a leader address this aspect of combat preparation? I am not suggesting a checklist approach or a recipe. I encourage young leaders to seek the advice of leaders who have, over the course of more than 10 years, prepared themselves, their families, and their Marines and their families for combat several times. Additionally, I would add one tool that helped my wife and me, which was the use of family war games. We tried to consider different chaotic events that might occur while one of us was gone. You might be laughing, imagining two Marine officers applying the Marine Corps planning process to family readiness. Laugh away. We took a chapter right out of Marine Corps Warfighting Publication (MCWP) 5-1 and played out what the parent at home and the deployed parent would do.[6] Our scenarios ranged from personal to logistical, including dealing with a sick child or death of one of our parents or considering a $20,000 maintenance issue for the car or home. We talked through what we would do, how we would pay the bills, and if and how we would travel as well as other aspects of how to handle the crisis. Then, we talked to our extended families about the same. For example, if one of my parents had died, I would have had to come home. However, we let everybody know that if one of my in-laws died, that I would not be there.

When my wife deployed to Afghanistan a year later and I stayed home with the kids, we dusted off the war games and reviewed them again. This activity forced us to do other things, such as update our will, powers of attorney, medical powers of attorney, and child custodial considerations in the event of catastrophe.

One additional benefit of these rehearsals was that I felt better prepared to do the same for my Marines who needed to have these discussions with their spouses and extended families. Addressing the family in a meaningful way (not just checking off the boxes on a predeployment checklist) is, in my view, equally critical for good preparation as physical and professional readiness is for combat. In the best case, this preparation might prevent a situation from escalating by addressing it before the Marine deploys. Maybe it keeps a Marine in the fight instead of disrupting the individual

---

[6] This publication details how Marine units should prepare to execute a range of military operations. See *Marine Corps Planning Process*, MCWP 5-1 (Washington, DC: Headquarters Marine Corps, 2010).

and that individual's small unit. At the very least, this family war game preparation equips a leader with the background knowledge to deal with the emotional trauma faced by his or her Marines.

BALANCING INDIVIDUAL VERSUS TEAM EFFORTS

Finally, I will address what I called the individual effort break point for leaders. I have always been a hard worker. Pure effort is the universal equalizer for those of us granted less than a full scoop of talent. As a lieutenant, I could overcome my own weaknesses, and sometimes the weaknesses of others, by putting enough effort into any exercise or project. By simply coming in earlier and going home later, I could control and fix almost anything. As a first lieutenant and captain, I honed my skills of micromanagement until I could just about personally supervise everything in my sight picture.

My fitness reports were good because, not only was my performance good, but my bosses could clearly see I was committed. Pleased with myself, I went to AWS and then worked for two years at Marine Corps Systems Command. As an individual project officer and a small team leader, I was again able to overachieve on pure effort. Then, I attended Command and Staff College as a major before arriving at the MEU, certain I knew it all. This inflated vision of my own abilities was about to unravel. As a MEU logistics officer, I became responsible for multiple functional areas (transportation, supply, fiscal accounting, equipment maintenance, and health services), the collective material readiness of three different subunits, and the management of training while planning three to six months in advance. The MEU was too big for me to succeed by working harder.

In less than two weeks, as a major with 12 years of service, I was forced to accept what pilots culturally learn at their training squadron during the first week: pilots do not know anything about fixing airplanes; they have to trust their maintenance crew. Infantrymen learn the same basic lesson the first time they take a platoon to the field: they cannot be everywhere at once and have to trust their subordinates to act in their absence. The individual effort break point, where one realizes that only the team will succeed and no amount of individual effort can overcome the situation, requires a feeling of complete dependence. Once the break point lesson is learned, officers focus on the skills of coordinating the efforts of subordinates and working with peers.

For me, this understanding came very late. I was a major on my first deployment with the MEU in 2006. This was the first time that the conditions of pure breadth and scope of responsibility in a MEU combined with the degree to which logistical functional areas were actually practiced instead of studied, and the extremely rapid exposure of one's folly when he or she talks without knowing taught me the lesson of the individual break point. The lesson unfolded for me through interaction with my subordinates in various areas of responsibility. I remember three conversations in particular that shaped my realization about the individual effort break point.

The first conversation was with my ammunition chief. Real ammunition, and a lot of it, had to be shipped to a specific location, stored, issued, fired, recovered, returned, and accounted for. A gunnery sergeant explained to me how this would work. I realized he was speaking in a language of ammunition acronyms that I did not know, and I did not have time to learn them even if I stayed in the office for a week straight. I had to trust that he knew what he was doing. The second enlightening conversation was with my MEU supply officer. He briefed me on all the stuff we, as a command element, had to buy for the attaching subordinate units. I asked, "Why don't they buy it for themselves?" He then showed me the comparison of the cost as it would affect an infantry battalion's annual budget. Thus, I got my introduction to contracting. Again, I felt completely helpless that an officer who was junior to me by three or four years seemed three or four years ahead of me in this particular critical function. Finally, I had a conversation with a limited duty officer 0430, or mobility officer, who is generally the most valuable player of every MEU in my opinion. During this conversation, she used so many acronyms so quickly that the only thing I understood was "and so that's why, sir." Now, whether or not I should have been more capable when I took that job is a valid question, but irrelevant to my point. For the first time in my career, I was responsible for things that I did not know thoroughly. Furthermore, if those things were not handled professionally, methodically, and completely, the poor results would impact 2,000 people and be on view immediately and tangibly for all onlookers. No amount of effort or long work hours can replace the years of experience represented by the senior Marines in particular disciplines on a MEU staff.

These were watershed events for me. Humbling. True, but more importantly, I had to quickly reevaluate how I approached my relationship with others up the chain, especially my commander, and how I understood the relationships I had with 40 Marines on a MEU S-4 (logistics) staff. The hardest part for me was going to the boss with a plan created by subordinates, because they were the ones with the technical competence to come up with the plan, and then selling it to him without completely understanding how all the pieces would fit together and if the plan would actually work. And sometimes, the plan did not work out; what then?

Well, a leader can blame his subordinate and keep himself or herself looking shiny. As much as every leadership school Marines have ever attended teaches exactly the opposite, too often I have seen and have been tempted to blame the Marine whose job it was to know that discipline to reduce my exposure to the wrath of the commander. However, that only works once. The next time a subordinate ensures that whatever plan goes forward belongs to someone above him to shift the blame if the plan goes south. If a leader wants a team that works, he has to take all the blame and none of the credit and avoid that mistake in the future. Repeating that cycle as often as possible, he will soon find his folks have done everything in new ways and made fewer mistakes. Furthermore, as the leader's competence increases, his role as translator to the commanding officer and subordinate unit is revealed.

If a logistics officer with 12 years of experience does not know much about the details of ammunition, purchasing/contracting, and strategic mobility, is it reasonable to assume your peers in other MOSs will? In deploying the force to Afghanistan, very experienced and, in my opinion, competent logistics staff members were pushed in their particular fields to the very limits of their own professional expertise. Additionally, we were geographically separated, but we had to work together; we had to perform as an orchestra trying to play a tune everyone recognized even though the musicians were in different venues. Clearly, any attempt to micromanage this effort would have resulted in abject failure. The trick, then, was to learn the control lesson early in one's career as I described pilots and grunts do inherently. For logisticians, some do learn it inherently, especially truck platoon commanders and engineers. However, droves of logistics and supply officers spend their formative years learning the very specific tools of their technical trades, and we rightly laud these officers for being experts in their niche fields (e.g., accounting, supply administration, maintenance operations, food service, etc.). Rightly so, the Marine Corps needs officers who understand these fields. The dilemma is that, in the years required to become a technical expert, the general skills of leadership and dependence on others atrophy.

In addition to spending time in operational jobs that train technical skills like supply, finance, and maintenance management, these officers must seek the opportunity to be in charge of something that requires dependence on other Marines in separate geographical locations. You can be a port-a-john platoon commander if that is what it takes (before taking the rank of captain). Being in charge of an indoor MOS section, such as a maintenance manager or an assistant logistics officer, is not the same thing.

I learned to trust and coordinate efforts of Marines who knew their disciplines better than I did. And like all skills, the ability to coordinate disparate technical efforts without being an expert in any of them is painful to learn the more senior a leader becomes. The earlier the lesson takes root, the better an officer is prepared for the ranks of major and beyond when technical skills are less important than an ability to derive the most from a team. Logistics discussions, even within combat experiences, are often as interesting as watching a herd of turtles run the steeplechase. My own experiences—less than revealing some previously undiscovered truism about combat—go far to reinforce the lessons of the generation that preceded mine. First, to be expeditionary, we have to practice movement and sustainment. During peacetime, the MEUs are the only tangible and funded laboratory for this training, which is so critical to our service identity. Furthermore, Marines need to train by actually operating while remaining small enough that real warfighting function integration happens even at the lieutenant and captain levels. Second, it does not matter what job a Marine does, combat will demand the very limits of his or her mental and physical capacity in all regards. So be ready. In my opinion, the trickiest of these personal preparations is the hardest one to measure and the easiest one to take a Marine out of the fight. Pay particular attention to family. Finally, mission success is based on teamwork. Effective teamwork is built on trust and personal relationships. No amount of hard work or Power-Point slides or computer system monitoring will replace the dynamic of people trusting you and you trusting people.

REFLECTION POINTS

- How important is real-world training for Marines and servicemembers in the administrative or logistics MOSs? How important is it for a leader in those MOSs to understand how expectations change once units are deployed to combat zones?

- How can military units use Marines' prior deployment experiences to better prepare military families for future deployment cycles? How could Lieutenant Colonel Carpenter's war game suggestion be integrated into units' deployment checklists for families?

- Lieutenant Colonel Carpenter sums up his experience with the 24th MEU as being reliant on trust in his subordinates' abilities to accomplish the unit's mission. How important is the element of trust for a military officer during a combat deployment? How might the leadership's management style cause a unit's mission to succeed or to suffer?

CHAPTER ELEVEN

A PERSPECTIVE ON LEADERSHIP
ATTRIBUTES IN COMBAT
MAJOR BENJAMIN P. WAGNER

Major Benjamin P. Wagner is a career infantry Marine. He enlisted in 1995 and was commissioned as an officer in 2002. Upon graduation from TBS and IOC in 2003, he was assigned to 2d Battalion, 1st Marines; he deployed twice with the unit to Iraq in 2004 and 2005 as a rifle platoon and mortar platoon commander. From 2006 to 2009, Major Wagner served as an instructor at TBS and IOC. Upon graduation from EWS in 2010, then-Captain Wagner deployed to Afghanistan as the commanding officer of Charlie Company, 1st Battalion, 8th Marines. Following company command, Major Wagner served as the operations officer for 2d Battalion, 2d Marines, and Special Purpose MAGTF-Black Sea Rotational Force before reporting to EWS as a faculty advisor.

In the opening pages of *A Moveable Feast*, Ernest Hemingway states, "After writing a story, I was always empty and both sad and happy . . . and I was sure this was a very good story although I would not know truly how good until I read it over the next day."[1] For me, when I am alone with my experiences and memories, this sentiment evokes the emotions associated with leading Marines in combat. I sincerely believe that any warrior who has faced the challenges of combat—the uncertainty, the confidence, the elation of success, and the dark, secret mistakes of which only he or she is aware—understands the conflicting emotions that Hemingway described. Yet, combat is real, it is not a story or fiction, but my memories of leading Marines into harm's way and making decisions that directly impacted their very existence have left me both happy and sad at the same time. And while we pound our chests and like to believe we did things right, combat and the price paid to win our nation's battles can leave honest warriors empty and reliant on the hope that when someone reads about our endeavors, that person will say that they were truly very good. Much has been written over the last decade about the fighting in Iraq and Afghanistan. I wonder, even now, why it is important or even necessary to write about my preparation for and experiences in combat. Perhaps, it is because I need to convince myself that what we have done is truly good. The reality is that only time will tell.

As a young enlisted Marine in the mid-1990s, I grew up in the shadow of Operation Desert Storm in Iraq and Operation Restore Hope in Somalia.[2] My squad leaders and platoon sergeants had combat action ribbons and stories of fighting in the desert of Kuwait and the streets of Mogadishu, Somalia. I also studied historical accounts of American involvement in World War II and Vietnam. I dreamed of the opportunity to test myself against opponents who would challenge me on the battlefield in the most visceral ways. I had no idea what it would really be like to deploy and fight, but I wanted to know. I wanted to feel it. It took almost 10 years before I had the opportunity.

---

[1] Ernest Hemingway, *A Moveable Feast* (New York: Scribner, 1964), chapter 1.

[2] Operation Restore Hope was the code name for the U.S. initiative, or United Task Force-Somalia, which carried out the United Nation's resolution for intervention in Somalia (1992–93). See Col Dennis P. Mroczkowski (USMCR, Ret), *Restoring Hope: In Somalia with the United Task Force, 1992–1993* (Washington, DC: Marine Corps History Division, 2005).

In 2003, while I was a student at the IOC, the U.S.-led Coalition invaded Iraq and so began more than eight years of Operation Iraqi Freedom. Those of us fighting the Centralians in the Quantico highlands sincerely believed that we had missed the big one.[3] We believed that we had missed the opportunity to test ourselves in a true maneuver warfare experience.[4] What we did not know then, but is obvious now, was that each one of us would get a chance to share in the truly sacred honor of leading America's chosen children into harm's way.

In September 2010, as commander of Charlie Company, 1st Battalion, 8th Marines, beginning my third combat deployment, my Marines, sailors, and I inserted into Now Zad District in Helmand Province, Afghanistan, to conduct a relief in place operation with a company from 1st Battalion, 2d Marines. We flew in from Camp Leatherneck on Sikorsky CH-53 Sea Stallion helicopters in the early morning hours just as the sun was rising over the high desert peaks surrounding Now Zad valley.[5] This district had seen very heavy fighting in 2008–9 when Marines from 2d Battalion, 7th Marines; 3d Battalion, 8th Marines; 3d Battalion, 4th Marines; and 2d Battalion, 3d Marines, had wrested control away from the Taliban and helped to reinstall the district governor and his team. Throughout 2010, the Marines of 1st Battalion, 2d Marines, worked hard to transition the district from one dominated by kinetic fighting to one of reconstruction and growth. Our mission was, as stated to me personally by the commanding general of 1st MarDiv (Forward), Brigadier General Joseph L. Osterman, to provide a hard shoulder against the Taliban strongholds to the north and to facilitate growth in the district and to tie its government and economy to the provincial capital in Lashkar Gah, Afghanistan.

---

[3] The Centralian Revolutionary Force is a fictional opponent used in field exercises at MCB Quantico.

[4] Maneuver warfare is explained in *Warfighting*, Marine Corps Doctrinal Publication 1 (MCDP 1), and is described as an attack on the enemy's system instead of killing enough of the enemy through attrition to force an end to the war.

[5] Camp Leatherneck, located in Helmand Province, was the base for Marine Corps operations in Afghanistan.

As a captain with fewer than 10 years in service as an officer, I was charged with leading more than 275 Marines, sailors, soldiers, and State Department civilians spread across more than 140 square kilometers with 13 fixed positions. Additionally, I was responsible for partnering with and training an Afghan National Army company as well as a district police force of more than 60 policemen. Finally, I was the principal liaison, coach, and advisor for the district governor and his team of government officials responsible for the daily leadership of the population in the district. While this may seem to some as a daunting task, I was fully prepared and able to handle these complex challenges because of the training and education I received in both formal schooling and, more importantly, from the lessons I had learned from my peers and senior leaders.

The challenges of leadership and command in combat are dynamic and complex. It is impossible to write one book or one chapter that captures all the lessons one can learn in combat. Truth be told, the lessons observed in combat do not truly become lessons learned until one has the opportunity to put them into practice while holding positions of leadership and responsibility in the future when one can apply those lessons. My personal experiences afforded me the opportunity to lead, teach, train, and prepare Marines for three combat deployments to both Iraq and Afghanistan. Additionally, I had the distinct honor and privilege to teach Marine lieutenants at TBS and the IOC as they prepared for combat deployments. From 2003 to 2013, my sole focus as a professional infantry officer was on training and preparing for and executing combat operations at the tactical level. My peers and I studied, discussed, and practiced doctrine and tactics, techniques, and procedures that we knew to work in the complex environments in which we would deploy and fight. We used both formal and informal methods of sharing ideas, lessons, and experiences in a constant effort to refine our ability to make sound decisions that would accomplish our assigned tasks and give our subordinate Marines and sailors the greatest chances to return home safely. We learned from one another and challenged ourselves mentally and physically to reach the highest levels of preparation for success in combat.

Now that I was being sent to Now Zad, I had a chance to bring all that experience to Afghanistan. As I would be reminded time and again during my seven months in Now Zad, the lessons I had learned—while fighting in Iraq as a lieutenant, teaching as a young captain and instructor at TBS, and studying as a student at EWS—prepared me for the challenges of company command and the decisions I would have to make in Now Zad. Those lessons learned through personal experience in combat, however, proved to be the most valuable when making tactical decisions. I needed to lead with the compassion required to inspire and motivate subordinates so they could overcome challenges and obstacles on a daily basis while continually being faced with the threat of enemy fire and IEDs.

Preparing for combat is what we do as Marines. We are constantly honing our skills and attempting to build training opportunities that test our brains and bodies to handle the rigors of combat. It is our obligation, as leaders, to draw on the experiences with which we are faced to develop an ability to replicate these myriad challenges in training to prepare the minds and bodies of our Marines for future situations. During the next few years, it will be a great responsibility for the Corps to capture the lessons we have learned in the last decade and to codify them into doctrine, to instill them in our junior Marines, and to ensure that we do not forget the valuable knowledge we have gained as individuals and as an institution to be highly capable in the complex and dynamic environments of modern-day conflict.

Preparing for the next war is not going to be easy. Unfortunately, for most Marines, this means fighting the last war. Buried in a small clearing on MCB Quantico, one can still see the remnants of a mock village established to teach Marine lieutenants how to operate in Vietnam. Urban training centers built in the 1980s are modeled after European towns. Our bases and stations today are filled with training venues, which mimic Middle Eastern environments. While funding is available during a conflict, our institution focuses on training where it knows we are going to fight. It is rare to find a unit, no matter how large or small, that is properly oriented on the what-ifs of a future battlefield. Analysts at think tanks like to put together working groups and host conferences focused on future strategies or operations, but at the tactical level, the vast majority of Marines draw from past experiences to build scenarios and environments with which they are comfortable. As an enlisted Marine in the 1990s, this meant patrolling for days on end because that was what my platoon sergeants had been taught by their platoon sergeants. Since the straight-leg infantrymen did very little face-to-face fighting during Desert Storm, there was also the mind-set that smart weapons and stealth aircraft were going to win our battles as they had there and in Kosovo at the end of the 1990s. Moreover, the lessons of Mogadishu did not adequately make their way into the battle drills and standard operating procedures of the Marine Corps infantry platoon. We cannot let the lessons we have learned fall by the wayside as the United States draws down from combat operations in Afghanistan. As leaders, we are compelled to ensure the retention of that knowledge and experience to continue refining our warfighting capability to maintain our edge as we continually prepare for future conflict.

As I reflect on my experiences, the lessons I have learned and those I have been taught, I have come to the opinion that there are four basic attributes a leader must provide in combat. The first, and to me the most important, is technical and tactical competency. There is no excuse for mediocrity in the professional application of warrior skills. Far too many leaders let their warfighting prowess atrophy in favor of thinking big thoughts, and they simply do not study and practice the most basic and fundamental skills we expect junior Marines and sailors to learn. If I failed to continue refining my abilities to apply the full range of current and valid tactics, techniques, and procedures, I believe that I would not fully employ my subordinates to the maximum extent possible. Our junior leaders will only properly apply the art and science of warfare if held accountable by their experienced and proven senior leaders. Those leaders who most positively influenced me in combat were those who demonstrated the abilities to both think at levels appropriate to their ranks and to understand what their subordinates were doing and when they were tactically wrong.

While serving in Afghanistan, my battalion commander regularly gathered the company commanders together once a month to discuss tactics employed and possible lessons to ensure that we were learning from one another and taking full advantage of every opportunity to win. After approximately three months in-country, he gathered us together at his command post to discuss search and attack tactics. The period of instruction he gave, along with numerous examples from history and our current operations, demonstrated a keen understanding of the battlefield and the responses of the enemy to our operations. He provided several examples of how best to employ scout snipers, observation platforms, information sensors, and boots on the ground to stay ahead of the enemy and make him react to our maneuvers. His desire to ensure we learned from one another and that we understood what he was seeing proved incredibly important to me as a company commander. I was able to better shape the battlefield, setting conditions in favor of my Marines on patrol. I ensured my company command post was reporting better, thereby keeping my higher headquarters aware of the tactical situation so that the full weight of the battalion could be brought to bear in a particular situation. This example of a technically and tactically proficient commander who knew his craft left a lasting impression on me and helped me develop as a company commander.

The other three attributes that I believe a leader must provide in combat are more ethereal: grace, dignity, and good judgment. I will never claim, however, that these are original thoughts of mine. Over time and through a multitude of personal experiences, I have come to value these attributes above others as vital to my success as a combat leader. I believe these characteristics are often not adequately understood by those preparing to lead and fight in combat. I did not fully understand how these attributes summarized the role and responsibilities of a combat leader until after my deployment as a company commander in Afghanistan. I came to realize that I needed to be a combat leader who served as a role model, a teacher, an inspiration, and a judicious decision maker to my fellow Marines and sailors in my company. At no other time would I have such complete influence over the bodies, souls, and minds of those serving with me in combat. The weight of the responsibility and the impact I had on my subordinates in combat demanded that, at all times, my thoughts, words, and actions were tempered by grace, dignity, and good judgment.

Grace is not a commonly used word in the Marine Corps and, I would submit, rarely understood or applied to the actions of Marines. We pride ourselves on being aggressive. We seek ways to set ourselves apart from other Services by being harder. This is true and is in the very nature of who we are as Marines. Yet, as a leader in combat, I believe that I must comport myself with the attitude of gentility and favor that enable me to have the most positive and lasting impact on my subordinates. Grace connotes an attitude of clemency, of patience and understanding. It requires a leader to understand the challenges faced by fellow Marines and an ability to put oneself in the same shoes as those of his or her subordinates. A commonly used phrase, "grace under pressure," implies an ability to smoothly handle complex challenges or to demonstrate maturity and calm in difficult situations. There are both secular and religious definitions of grace. I submit that a combat leader must be both aware of and able to apply both definitions. The ability of a combat leader to effectively connect with a subordinate when discussing complex or challenging events must be coupled with the ability to demonstrate patience and understanding. The concept of grace extends to those situations when a subordinate might not have earned the patience and understanding of the combat leader. These are perhaps the most challenging leadership opportunities. A combat leader can easily cause long-lasting damage to a subordinate by not seizing an opportunity to demonstrate grace, both in the ability to look at a situation through the lens of subordinates or to be patient and understanding of them.

Combat, as defined by military theorist Carl von Clausewitz and our own Marine Corps doctrine (MCDP 1) is chaotic. It is uncertain and ever changing and requires mental and emotional flexibility and adaptability. I believe that as a leader, I must learn to overcome initial responses and maintain a calm and collected attitude that will positively inspire my subordinates to ever-greater levels of achievement. This does not mean that I accept failure. It does not mean that I fail to hold my subordinates accountable for their actions in combat. It does mean, that as a combat leader, I look at every situation as unique and take into account the persons involved, the environment, the outside influences, and the challenges to a subordinate's decision making.

As a second lieutenant, I learned much about grace and leadership while serving under then-Captain Douglas Zembiec in Echo Company, 2d Battalion, 1st Marines.[6] We deployed to Fallujah, Iraq, and fought and served there from March through October 2004. We were taking part in Operation Vigilant Resolve in response to, among other things, the murder and mutilation of four Blackwater security contractors on 31 March. Our battalion, along with three others, pushed into the city to regain control and solidify the hold by the legitimate government of Iraq. Despite the political unrest and back and forth at the highest levels, we were in a fight and no cease-fire was honored by the insurgents in the city. During these formidable seven months, I learned some of the most heart-wrenching lessons through personal experience.

On 26 April 2004, following a firefight during which a Marine in my platoon was killed, my company commander sought me out because he knew I was in need of some focused leadership. He knew that my virtual tank was running low and that I was frustrated with the continued losses in my platoon. It seemed as though my platoon in particular was continually getting into firefights that resulted in the wounding or death of wonderful, strong, young warriors. Due to the nature of urban fighting, the combat was close and personal; we could hear and see our enemies just feet away, at times, as they sought vantage points from which to fire on us.

On this particular day, the fighting had been very heavy. The regimental operations officer told me, some time later, that he had seen more than 100 insurgents move toward our platoon's position while watching the fight on a live feed from an unmanned aerial vehicle. Two Marines held the enemy from breaching into our strongpoint by dropping hand grenades over the side of the roof. Another Marine used his M249 squad automatic weapon from an exposed position in the street to cut down insurgents

---

[6] For more on Zembiec, see chapter 1.

moving in the open. While evacuating several of our wounded comrades, two Marines used their automatic weapons and an M203 grenade launcher to devastate insurgents firing on us from an adjacent building. Personal heroic action in this firefight resulted in my Marines receiving three Silver Stars, four Bronze Stars, and numerous other valor awards. Of the 24 Marines and corpsmen in my platoon, 11 were wounded on that day. It was a tough fight, and my brain and body were drained and tired.

Several years later, when I was a staff platoon commander at TBS, I began to fully understand what my company commander did for me following that fight. At that time, one of the required readings for the lieutenants was *Passion of Command* by Colonel Bryan P. McCoy. A short book with many valuable leadership and training lessons, there was one particular passage that speaks directly to what Doug Zembiec did for me on 26 April 2004. In the book, Colonel McCoy discusses the "well of fortitude."[7] Though he borrowed the idea, he did a very good job of explaining the role of a tactical leader to manage the well for his or her subordinates. My well was running dry, and I needed a more experienced leader to help me find a source of strength to overcome the challenges I was facing on a daily basis.

Doug told me that aggressive first and second lieutenants needed to be tempered by experienced captains and that experienced captains needed to be tempered by judicious lieutenant colonels. He let me know that I was making the right decisions and that my Marines were doing exactly what they were supposed to be doing. Moreover, we had trained the men well, and they were taking the fight to the enemy and effectively killing them in every engagement. He told me that casualties happen, and as long as I was employing my Marines and their weapons appropriately, I must be proud of their sacrifices and rejoice in the efforts they put forth to fight and win with dignity and honor. I would not fully understand the lesson he taught me that day until it was my place to fortify junior Marines and sailors in future combat deployments.

---

[7] Col Bryan P. McCoy, *The Passion of Command: The Moral Imperative of Leadership* (Quantico, VA: Marine Corps Association, 2007), 19.

Captain Zembiec demonstrated grace toward me in the attitude he took to understand the pressures I was facing and to reach out to me in a way that showed he could accurately grasp the complex challenges of being a young platoon commander. He did not tell me to suck it up or to be harder or that I was weak for worrying about casualties. He let me talk and he reinvigorated my soul, thereby making me a more effective combat leader. His example of grace instilled in me a greater confidence in his leadership and a better understanding of what he wanted from me. It made clear to me that his expectation was that I continue challenging the enemy, while pressuring the insurgents at each and every opportunity. He inspired me to continue pushing my Marines, but he did so with compassion and an understanding that communicated to me that he was fully aware of what we were doing and how that taxed our well of fortitude. Through his personal example, he inspired me to continue striving for greater levels of effectiveness and success in combat.

Another trait that I believe is vital for a combat leader is dignity. Dignity is a much more commonly used word in the Corps than grace, but I would submit it is not necessarily understood by all. I feel that I was able to demonstrate dignity as a combat leader by elevating my character and conducting myself in such a way as to clearly show an understanding for the gravity of a situation. As I have studied the concept of dignity and its many descriptions and definitions, I have come to believe that since there is a significant moral aspect of combat and the actions taken by combatants, I must also provide moral leadership to my subordinates. This is simple in its basic form, and I doubt that there is any member of the U.S. armed forces who would argue this point. It is important to recognize, though, that it is this moral aspect that requires dignity from a leader in combat. The concept of dignity that I propose a combat leader must demonstrate is tied to the constant awareness that subordinates, as humans and as the cherished children of our country, deserve a leader who never loses sight of their worth and value. As a leader in combat, I must understand that, at every moment of every day, there are Marines in my charge risking their lives to accomplish their assigned tasks. Every time Marines leave friendly lines, they are prepared to make contact with the enemy. I must be a leader who demonstrates dignity when interacting with these young warriors. In so doing, I can and will have a lasting impact on their preparation for combat and the way in which they conduct themselves under fire.

On 12 April 2004, my platoon was in a defensive position just inside the northern edge of Fallujah. On that particular evening, my platoon came under fire from insurgent forces to the south of our position. We were at stand to (ready for attack), just as I had been taught at the School of Infantry and TBS. The Marines were prepared for contact, and we got it. During the firefight, a friendly 81mm mortar round landed on my platoon's position, killing two Marines and a U.S. Army interpreter and wounding several other Marines. Because we were in an urban environment, we were not able to spread out as we would have in open terrain, and the channelizing effect on the shrapnel caused many more casualties than would normally be expected.

In 2004, I was a second lieutenant who had been in combat for just one month, and two of my Marines were dead. I can still recall the overwhelming pressure I felt that night. I can see the corpsmen and my platoon sergeant working on the wounded and still clearly feel one of the mortally wounded Marines squeezing my hand as I encouraged him to hold on. He pleaded with me to get him to the aid station. I remember what it was like to walk the lines that night after the fighting had died down. I remember what it was like to explain to the Marines on post that two of their brothers were dead. I remember what it felt like to sit in a dark room and put my head in my hands, to wonder what I could have done differently, replaying every detail over and over in my head and asking if I did anything wrong.

My platoon sergeant saw me in the middle of the night and saw that my uniform was covered in blood. He told me to get changed and to clean myself up. I was so incredibly tired, and the only thing I wanted to do was to lie down and rest my brain and my body. He told me that the Marines would be looking to me for strength and that it was my job and role to inspire confidence and to let them feed off of me for the energy to go on and for the calmness of spirit not to lose sight of their mission. He told me that if they saw me clean and focused, conducting myself with dignity, they would understand what I expected of them. That experience not only made a significant impact on me, it also taught me a vital lesson about the fundamental role an experienced and capable staff NCO fills in teaching sound leadership; the wise platoon sergeant instructing the tired and emotionally drained lieutenant. I did not learn that in a lecture at TBS or in a field event at IOC. My platoon sergeant taught me because he understood what a leader in combat must provide to his or her subordinates.

In 2010, as a company commander with 1st Battalion, 8th Marines, in Afghanistan, conducting myself with dignity meant demonstrating to my Marines and sailors that I understood what they were doing and the challenges they were facing each and every time they left the wire. I was not on every patrol, but it was my place to communicate to them through my actions that I was aware of what they were doing and did not take for granted the efforts they put forth to maintain security throughout the Now Zad District. I took every opportunity to see and talk to Marines before they left friendly lines to conduct patrols. I would observe the patrol leaders conducting their precombat checks and inspections; I would talk to the Marines, look them in the eye, and listen to their concerns.

I am convinced that leaders who demonstrate dignity in a combat situation sleep very little. I was constantly concerned that each and every decision I made was right, tempered by experience and knowledge and focused on providing my subordinates the greatest chance of success in each and every assigned task. It does not mean that I became averse to risk and was then unwilling to put my subordinates in harm's way. Quite the opposite, it means that I was obligated to take the time to prepare, study, examine, and think deeply on how to employ my assigned people, weapons, and assets to the maximum extent possible to defeat the enemy soundly with the minimal cost possible.

Regretfully, with great examples of leaders acting with dignity, I have also observed poor examples by leaders of all ranks in combat. Those who stay inside the wire do not share the challenges of their subordinates and do not walk the lines at night and talk to the Marines on post; they all demonstrate a lack of dignity that is easily recognized by their subordinates. Those who carelessly order their subordinates to do foolish things in combat such as unnecessary movements, presence patrols, or other tactically unsound actions without a clear purpose communicate a lack of understanding for what our Marines and sailors do each day in combat. Additionally, those who cannot talk to their Marines, who cannot listen and communicate well the intent behind decisions, fail to conduct themselves with dignity. Perhaps one of the best books written on the human dimension of leading in combat, Anton Myrer's epic novel, *Once an Eagle* (1968) perfectly captures the essence of a combat leader who personifies dignity in combat with the fictional character of Sam Damon who is contrasted with a leader who does not, in the character of Courtney Massengale. I strongly recommend that anyone interested in further examining the importance of a dignified combat leader read and reread this book.

Finally, my Marines and sailors wanted a leader who made decisions. They wanted those decisions to be timely and wanted them to be clearly articulated. I could not waffle when making decisions or wait until I had every last piece of information as this would be slow and expose my subordinates to unnecessary risk. The ability to make effective decisions is not a trait with which leaders are born. The skill is learned over time and through experience. As a combat leader, I had the obligation to both make good decisions and also to teach my subordinates how to make good decisions. The ability to do both was a vital requirement for me to be a capable and effective combat leader. Failure to do so would inevitably result in causing friction among and harm to my subordinates. This adaptive decision-making style is frequently referenced in current leadership development studies and is recognized as a significantly challenging aspect to how the Marine Corps educates and trains future leaders.[8]

---

[8] For more on training future military leaders, see *U.S. Marine Corps Small Unit Decision Making: January 2011 Workshop Final Report* (Quantico, VA: Marine Corps Training and Education Command, 2011); LtCol William J. Cojocar (USA, Ret), "Adaptive Leadership in the Military Decision Making Process," in "Mission Command Symposium," *Military Review,* special issue (June 2012); and Michael Useem, "Four Lessons in Adaptive Leadership," *Harvard Business Review*, November 2010.

Like so many other Marines, I have observed both good and bad examples of decision making in combat. Due to the nature of combat, decisions are made constantly by leaders at all levels. As a company commander, I was proud to watch my lieutenants make excellent decisions on a regular basis in incredibly challenging and complex situations. Decision making in a counterinsurgency environment puts an incredible amount of stress on junior leaders and demands their constant situational awareness and focus. They must continually assess their environment and their understanding of commanders' intent and make decisions that are adaptive in nature. They must recognize patterns and distinguish anomalies. They must make split-second judgments and be equally comfortable acting aggressively or demonstrating extreme patience as a situation develops.

Two particular examples of effective and appropriate decision making stand out to me from my time as a company commander. One month after my unit's arrival in Afghanistan, a mounted patrol was responding to the identification of an IED on a road in southern Now Zad. According to our approved procedures, the Marines on patrol appropriately established a cordon around the site to prevent the civilian population from wandering near the device while the explosive ordnance disposal team was prosecuting the IED. The patrol leader and all other small unit leaders were doing exactly as we had rehearsed time and again during our predeployment training. They knew what they were supposed to do tactically and did it very well. Before the device could be destroyed, however, a vehicle traveling at a high rate of speed was observed approaching the cordon from the north. A young lance corporal manning a vehicle-mounted machine gun observed the vehicle and reported its approach to his squad leader. We refer to this type of situation as an escalation of force event. It is perhaps one of the most challenging situations in which a young Marine can find himself or herself and requires split-second decision making, accuracy, and decisiveness. As the vehicle proceeded beyond the signs posted directing it to stay clear of the cordon and continued toward the lance corporal's vehicle, the driver demonstrated no intention of stopping. In seconds, the Marine rapidly employed a flare and when the vehicle failed to slow or stop he reached for his M4 carbine rifle (not the mounted machine gun in front of him) and fired a single, well-aimed shot into the engine block of the vehicle causing it to stop immediately just 150 feet from the cordon.

This rapid and decisive example of a young Marine making a series of sound decisions is exactly what I hoped my Marines would do when placed in a challenging situation. The lance corporal made every attempt to stop the vehicle, and when he determined that fire was warranted, he chose to use his carbine instead of his heavier machine gun, thereby limiting collateral damage or unnecessary harm to potential innocent civilians. In fact, the car turned out to contain five unarmed civilians who had no intention to harm the Marines, but were simply unaware that their actions were perceived as a threat. The Marine's actions that day gained the praise of the police chief and district governor. The Marine demonstrated dignity and sound decision-making in the appropriate application of force to accomplish his mission and secure the safety of his fellow Marines. Even today, this memory remains with me as an example of everything done exactly right by a young Marine.

This was not, however, a skill with which the lance corporal was born. Through years of operating in a complex COIN environment, we have learned the dangers in causing collateral damage or unintended harm to civilians. We have learned that the cost of unnecessarily injuring a civilian (even one who is doing wrong) is significant and can cause serious second- and third-order effects. Because of this, we have developed detailed training events to challenge our young Marines in exactly this way—to train their minds and bodies to make hard and good decisions in a time-compressed environment. We practice this particular type of situation over and over to formulate conditions so our Marines will gain the ability to distinguish right from wrong under pressure and to recognize threats and patterns. Adaptive combat decision making includes demonstrating patience in the face of uncertainty. Being an effective leader in combat does not mean always being the first to decide to act. It means that the leader must act correctly and at the right time. A leader who rushes to decisions leads with his chin. The rushed and uninformed leader is like a boxer who expends all of his energy in the first round of a fight, swinging randomly with no plan for how to defeat his opponent.

In the last week of my deployment to Afghanistan, I observed patient, adaptive decision making in one of my platoon commanders. We had recently helped establish a police checkpoint in a hotly contested portion of the district along a known enemy infiltration route. There were numerous indications that the enemy intended to challenge this new position, and in response, under the cover of darkness, the lieutenant led a patrol to reinforce the police position. He got his Marines to the checkpoint and immediately had them move inside the structure and remain under cover in an effort to hide their presence from any enemy observers. Nothing happened that night or on the next day. He and I spoke on the radio, and I asked him what he wanted to do. He spoke with his subordinate leaders and thought for a while before telling me that he wanted to stay at the position for another night. Near sundown, the enemy began firing on the police position. I anticipated that the Marines would return fire, but the lieutenant, demonstrating a keen understanding of the enemy tactics and excellent decision making, kept his Marines under cover. He recognized quickly that the enemy fire was a probing attack, an attempt to get the police to expend ammunition (which they readily did). The platoon commander sincerely believed that the enemy would attack in force later in the night when they assumed the police were at their lowest level of awareness and preparation.

Just after midnight, the enemy did exactly what the lieutenant expected. They used RPGs and machine guns to conduct a coordinated assault on the police checkpoint. At this time, the lieutenant unmasked his Marines, and their fire devastated the enemy assault resulting in three enemies killed in action and no injuries for the Marines or Afghan police. The lieutenant demonstrated an adaptive decision-making ability, which is exactly what we must desire, develop, and expect of a combat leader. He recognized patterns, understood the enemy and the environment, correctly estimated the skills and capabilities of his subordinates, and knew what he could expect of his Marines. He was confident in his assessment and knew that I would trust him to carry out the decisions he made. He had learned how to think and make decisions. He had created an environment where his subordinate Marines trusted his instincts and were confident that his decisions would carry the day. This climate of attitude, confidence, and trust takes time to foster and a significant amount of effort for a leader to create. It requires long hours of education and training and the support of senior leaders. A combat leader who can successfully balance both aggression and patience has a significantly higher chance to succeed than one who rushes to decisions or does not adequately take into account the various facets of each situation.

Grace, dignity, and superior decision-making abilities take time and effort to develop. There are various means by which one can learn these traits, but above all, they take practice. None of these lessons are unique and none are different from what other leaders have observed in combat. They were, however, lessons that cannot be fully learned from a book or taught by an instructor at a formal school. They cannot be adequately taught in the field or in the lecture hall. These are the lessons that the combat leader is obliged to remember and for which he or she must prepare. These are the lessons that a leader with combat experience must ensure he or she teaches to future subordinates. I am convinced that a leader must spend vast amounts of time reflecting on those combat experiences, revisiting them time and again as he or she gets older to reflect on and allow them to continually refine his or her character. Combat leaves an indelible mark on one's soul. The lessons and memories are, as Hemingway once called his reflections on his time in Paris, "a moveable feast."[9] The lessons and memories remain with good leaders for the rest of their lives wherever they go and feed them well as they work so hard to educate the next batch of men and women who will be thrown into the fray.

REFLECTION POINTS

- Several authors have presented descriptions on the value of leadership as Major Wagner has done. Which approach toward leadership do you

---

[9] Hemingway, *A Moveable Feast.*

think is the best? Why do you think each account offers different attributes for good leadership?

• Have you known a leader (military or civilian) who made an impact on you and others? What are the characteristics of that leader and how do those traits compare with those presented in this book?

• Would you agree or disagree with the author that grace is a key characteristic for a combat leader? Does this characteristic apply only to leadership in combat?

REFERENCES AND FURTHER READING

Armstrong, Keith, Suzanne Best, and Paula Domenici. *Courage after Fire: Coping Strategies for Troops Returning from Iraq and Afghanistan and Their Families*. Berkeley, CA: Ulysses, 2006.

Associated Press. "Marine Lance Cpl. Andrew S. Dang." *Military Times* Honor the Fallen database.

Associated Press. "Marine Lance Cpl. William J. Wiscowiche." *Military Times* Honor the Fallen database.

Boudreau, Tyler E. *Packing Inferno: The Unmaking of a Marine*. Port Townsend, WA: Feral House, 2008.

Camp, Dick. *Operation Phantom Fury: The Assault and Capture of Fallujah, Iraq*. Minneapolis, MN: Zenith Press, 2009.

Campbell, Donovan. *Joker One: A Marine Platoon's Story of Courage, Sacrifice, and Brotherhood*. New York: Random House, 2009.

Caputo, Philip. *A Rumor of War*. New York: Owl Books, 1996.

Carroll, Andrew, ed. *Operation Homecoming: Iraq, Afghanistan, and the Home Front, in the Words of U.S. Troops and Their Families*. Updated ed. Chicago: University of Chicago Press, 2008.

Chan, Sewell. "U.S. Civilians Mutilated in Iraq Attack," *Washington Post*, 1 April 2004, https://www.washingtonpost.com/archive/politics/2004/04/01/us-civilians-mutilated-in-iraq-attack/1c9a38a8-2570-4814-9850-0e6e5a766922/.

Commander and Staff, Marine Wing Support Group 37. "No FARP Too Far!," in *U.S. Marines in Iraq, 2003: Anthology and Annotated Bibliography*, comp. Maj Christopher M. Kennedy et al. Washington, DC: Marine Corps History Division, 2006.

Daly, Thomas P. *Rage Company: A Marine's Baptism by Fire*. Hoboken, NJ: Wiley, 2010.

Estes, LtCol Kenneth W. (Ret). *U.S. Marines in Iraq, 2004–2005: Into the Fray*. Washington, DC: Marine Corps History Division, 2011.

Fick, Nathaniel. *One Bullet Away: The Making of a Marine Officer*. Boston: Houghton Mifflin, 2005.

Gray, J. Glenn. *The Warriors: Reflections on Men in Battle*. Lincoln, NE: Bison Books, 1998.

Gregson, LtGen Wallace C. "I Marine Expeditionary Force Summary of Action," in *U.S. Marines in Iraq, 2003: Anthology and Annotated Bibliography*, comp. Maj Christopher M. Kennedy et al. Washington, DC: Marine Corps History Division, 2006.

Holmes, Richard. *Acts of War: The Behavior of Men in Battle*. New York: Free Press, 1985.

Holmstedt, Kirsten. *Band of Sisters: American Women at War in Iraq*. Mechanicsburg, PA: Stackpole Books, 2007.

———. *The Girls Come Marching Home: Stories of Women Warriors Returning from the War in Iraq*. Mechanicsburg, PA: Stackpole Books, 2009.

Junger, Sebastian. *War*. New York: Twelve, 2011.

Kasal, Brad, and Nathaniel R. Helms. *My Men Are My Heroes: The Brad Kasal Story*. Des Moines, IA: Meredith Brooks, 2007.

Keener, Michelle. *Shared Courage: A Marine Wife's Story of Strength and Service*. St. Paul, MN: Zenith Press, 2007.

Klay, Phil. *Redeployment*. New York: Penguin Press, 2014.

Kummer, Maj David W., comp. *U.S. Marines in Afghanistan, 2001–2009: Anthology and Annotated Bibliography*. Quantico, VA: Marine Corps History Division, 2015.

Livingston, Gary. *Fallujah, with Honor: First Battalion, Eighth Marine's Role in Operation Phantom Fury*. North Topsail Beach, NC: Caisson Press, 2006.

Marine Corps. *Leading Marines*, MCWP 6-11. Washington, DC: Headquarters Marine Corps, 1995.

Marine Corps. *Small Wars Manual*, NAVNC 2890. Washington, DC: Headquarters Marine Corps, 1940.

Marlantes, Karl. *What It Is Like to Go to War*. New York: Atlantic Monthly Press, 2011.

McClellan, Maj Edwin N. *The Unites States Marine Corps in the World War*. Revised ed. Quantico, VA: Marine Corps History Division, 2014.

McCoy, Col B. P. *The Passion of Command: The Moral Imperative of Leadership*. Quantico, VA: Marine Corps Association, 2007.

McWilliams, Timothy S. *U.S. Marines in Battle: Fallujah, November–December 2004*. With Nicholas J. Schlosser. Quantico, VA: Marine Corps History Division, 2014.

Mihocko, LtCol Melissa D. *U.S. Marines in Iraq, 2003: Combat Service Support During Operation Iraqi Freedom*. Washington, DC: Marine Corps History Division, 2011.

Morse, Dan. "Salute to a Memorable Marine," *Washington Post*, 17 May 2007, http://www.washingtonpost.com/wp-dyn/content/article/2007/05/16/AR2007051602860.html.

Myrer, Anton. *Once an Eagle: A Novel*. New York: Harper Collins, 2000.

O'Brien, Tim. *The Things They Carried*. New York: Broadway Books, 1998.

O'Donnell, Patrick K. *We Were One: Shoulder to Shoulder with the Marines Who Took Fallujah*. Cambridge, MA: Da Capo, 2006.

Pettegrew, John. *Light It Up: The Marine Eye for Battle in the War for Iraq*. Baltimore, MD: Johns Hopkins University Press, 2015.

Reynolds, Col Nicholas E. *U.S. Marines in Iraq, 2003: Basrah, Baghdad, and Beyond*. Washington, DC: Marine Corps History Division, 2007.

Stout, Jay A. *Hammer from Above: Marine Air Combat Over Iraq*. New York: Presidio Press, 2005.

West, Bing. *No True Glory: A Frontline Account of the Battle for Fallujah*. New York: Bantam Books, 2005.